VALUES, TECHNOLOGY AND WORK

SIJTHOFF & NOORDHOFF SERIES ON INFORMATION SYSTEMS

Consulting Editors:

Richard J. Welke
McMaster University
Hamilton, Ontario, Canada

Ronald K. Stamper
London School of Economics
and Political Science, London, U.K.

This Series will incorporate that on Information Systems Analysis and Design originated by the London School of Economics and Political Science

1. Design and Implementation of Computer-Based Information Systems, edited by N. Szyperski and E. Grochla

2. The Impact of Systems Change in Organisations, edited by N. Bjørn-Andersen, B. Hedberg, D. Mercer, E. Mumford, A. Solé

3. Values, Technology and Work, by E. Mumford

VALUES, TECHNOLOGY AND WORK

by

Enid Mumford
Professor of Organizational Behaviour,
Manchester Business School

1981

MARTINUS NIJHOFF PUBLISHERS
THE HAGUE / BOSTON / LONDON

Distributors:

for the United States and Canada

Kluwer Boston, Inc.
190 Old Derby Street
Hingham, MA 02043
USA

for all other countries

Kluwer Academic Publishers Group
Distribution Center
P.O. Box 322
3300 AH Dordrecht
The Netherlands

Library of Congress Catalog Card Number: 81-50361

ISBN 90-247-2562-3 (this volume)
ISBN 90-247-3003-1 (series)

PRINTED IN THE NETHERLANDS

Table of Contents

Preface

This book describes the experiences of four organizations who tried to introduce new computer systems in a humanistic manner so that human as well as business gains would be derived from the introduction of technology. All four paid a great deal of attention to identifying efficiency and job satisfaction needs and to designing the technical system and its surrounding organizational context in such a way that these needs could be effectively met. Nevertheless, as with all major change, the change process was difficult and demanding and considerable management skill and insight was required before successful systems were implemented.

The author set out to identify the extent to which the values of the different groups involved in the design process influenced the way in which computer systems were designed and implemented. She also wished to establish the extent to which the values of technical systems designers, user management and user clerks converged or diverged in the change process. It is hoped that the ideas set out here will contribute both to a greater theoretical understanding of the influences which affect technical change and to the practical design of humanistic computer systems. The research was carried out in three large government departments, two industrial firms and an international bank. Two of the government departments asked for their data to remain confidential and so these are not described in detail in the book.

The book is in twelve chapters. Chapter 1, the *Introduction*, describes the objectives of the research and discusses the origins of technical and rational values. Chapter 2 discusses *What are Values?* and provides a definition and analysis of values. Chapter 3 sets out the *Theoretical Tools* used for the analysis of values in the case study situation. These tools have been developed by the author although they are derived from the work of Talcott Parsons. Chapter 4 on *Systems Designers: Their Values and Philosophy* tests the hypothesis that the values of technical systems designers reflect a technical ethic. This hypothesis is not supported. Chapter 5 on *The Research Situations* describes how the research was carried out and provides a framework for analysis of the case studies. Chapters 6 and 7 discuss in detail the influence of values on the design of large scale computer systems in two industrial firms. Chapter 8 makes a similar analysis for a government department; chapter 9 does the same for an international bank, and chapter 10 draws some conclusions. Chapter 11 states conclusions on *Values and the Change Process* and shows how democratic and humanistic values produce different systems designs and consequences from technical-rational values. Chapter 12 presents the author's own value position and examines the problems of the future.

The author would like to thank the management of the firms, bank and government department involved for their kindness in cooperating in the study. Two of them did not wish their identity to be revealed and so the two firms and the bank have been given names which are not their true ones. The British Inland Revenue is correctly named throughout the text and Asbestos Ltd. is TAC Construction Ltd., a company in the Turner and Newall Group.

The author was enabled to undertake this research through the award of a Personal Research Grant by the British Social Science Research Council, to whom she is indebted.

1. Introduction

The Computer and Work Design Research Unit at Manchester Business School, with Enid Mumford as Director, has been working for many years in the research area of systems and work design and job satisfaction. An important objective of this group has been to influence management and unions to make the improvement of job satisfaction a major objective of any systems or work design process. Our studies have concentrated on the design of computer systems in offices although the Unit has also investigated the design of work in general, and carried out studies of shop floor technology, including group technology, transfer line technology in the automobile industry, and assembly line technology as it affects women workers.

Computer technology has played a major part in our work design research for a number of reasons. First, the computer acts as a major catalyst for change. Most computer systems require considerable rethinking of the organization of user departments and of the structure of user jobs; a new computer system therefore provides an excellent opportunity for introducing some job satisfaction objectives into the systems design process, these objectives being achieved through the way the new office organization and task structure are designed.

The earlier studies of the impact of computers on office organization and clerical task structures indicated that computer systems introduced relatively few technical constraints into the design of jobs in user departments, although some ancillary computer jobs such as punch operating, normally located in the computer department itself, were totally influenced by the computer. Despite this absence of technical constraints it had been found that the work structures associated with new computer systems were often segmented and routinised and had tight controls associated with them. It was clear that a factor other than the requirements of technology was influencing the design process and it was hypothesized that this was the philosophy and values of the groups responsible for systems design, particularly their perception of the competence of user staff and their beliefs on the best way of organizing work to achieve maximum efficiency. In order to understand and improve systems design by making it more receptive to human needs it seemed important to approach the subject from a new angle; that of gaining an understanding of the values, design principles and models of man used by systems designers, managers and others in the design of clerical computer systems. By "models of man" is meant the vision that the systems designer and his associates have of the needs, motivations and capabilities of the group who will receive and operate the new computer system. In addition to ascertaining the design philosophy of computer technologists it was also necessary to find out how this design philosophy was being translated into a concrete organizational form, particularly through the way in which work was organized and jobs were structured.

User groups will approve or disapprove of a new computer system on the basis of an assessment of the help it gives them in doing their jobs and its capability for providing them with, or at least not removing, the kind of work they want to do. Any evaluation of systems design policies and consequences therefore required information on the extent to which new computer systems were meeting the efficiency and job satisfaction needs of managers and staff in the user departments.

Previous discussions with systems designers suggested that some would have liked to have made greater provision for human needs when designing computer systems but felt that a humanistic approach would not fit well with the ideology and values of their companies. This suggested that design philosophies and practices must be considered in an organizational context and an attempt made to identify the kinds of moral pressures that top management either exerted on specialist groups, or were thought exert.

Enid Mumford was provided with an opportunity for investigating these new research areas through the award of a Personal Research Grant by the S.S.R.C. for the year 1974/75.

1.1. THE OBJECTIVE OF THE RESEARCH

The project started with a precise research objective derived from previous research carried out by the author and formulated in terms of questions which needed to be answered [Mumford 1972 and Mumford and Pettigrew 1975]. This research objective was to establish the nature and consequences of the design processes associated with a number of specific computer based administrative systems. This would involve describing the structures of work organization and tasks that emerged from these processes, identifying the extent to which these had been markedly influenced by particular design philosophies, and establishing whether such structures had encouraged or reduced staff feelings of job satisfaction with the task aspects of work. Through the study it was hoped to obtain answers to the three following questions.

1.   What were the values of the different groups in the research situations and to what extent were these values in harmony or conflict with each other? This examination of values would cover the following:

The values of the higher levels of management on the most efficient way of organizing work from either the management or employee point of view. These values could influence the approach to systems design of the computer specialists and line management.

The values of the technical systems designers on how computer systems should be designed. These values could be incorporated into the operating system

2

through the way in which work objectives were defined, tasks allocated and controls organized. It would be important to establish if the systems designers had set themselves any human goals such as an increase in *job satisfaction* and an improvement in the *quality of working life,* when designing these systems.

The values of the manager of the user department on the best way to organize work to achieve efficiency and motivation.

The values of the user department clerks on the most efficient and satisfying way of organizing work from their point of view.

2.   To what extent was the organization of work in the user departments, and the task mix of work groups and individual employees within these departments, influenced by, or a reflection of, these group values?

3.   To what extent were the computer based work systems in the research situations meeting the following needs:

a.   Technical and business efficiency as defined by management, computer specialists and the clerical labor force?

b.   Needs for job satisfaction and a good quality of working life as defined by clerks, management and computer specialists. Did the employees operating the systems in the user departments show a high level of satisfaction with their tasks and responsibilities? If a high level of job satisfaction was found to exist what features in the design of the system had contributed to this and was an absence of *human design goals* an important element in a system's unsatisfactory human consequences?

If efficiency and job satisfaction were viewed as deficient in any respect an attempt would be made to establish if the computer and associated work systems could have been designed in a different manner so as to reduce or avoid these problems. If alternative design strategies were available, were these considered and, if this was the case, why were they rejected?

The questions the research was set up to answer were formulated as hypotheses to be tested through data collected during the research. A first, key hypothesis, was that *industry, commerce and government are not at present getting clerical computer systems which are consciously designed to increase job satisfaction because the values which influence systems designers are not of a kind which motivate them to do this.* These values may be the systems designer's own individual or professional values or they may be top management or user values to which systems designers have to respond. An attempt will be made, in each research situation, to establish which of these value systems had the strongest impact on design outcomes. Previous research by the author had suggested that many systems are designed in terms of a vision of man and man's needs and abilities which is a product of the technical systems designers' own values, training and experience [Mumford 1972]. In a situation where the potential users of the system may lack time, EDP knowledge, and perhaps the motivation to become involved in the design process, the systems designer is left to create his own organizational reality and this may not coincide with the reality of the people who in the end have to work with these systems and operate them effectively.

3

A second hypothesis was that *systems designers, in an attempt to regulate and reduce the variety of the systems design process, build simple conceptual models of the design universe and the people located in it*. These models do not typically include human factors such as a desire for job satisfaction [Hedberg and Mumford 1975]. Such models assist the technical designer to bring order to the design process and when translated into objectives and procedures they determine the manner in which he approaches his work and the criteria which he uses for evaluating the success of his work. But again they are likely to be derived from his own value system and, if there are no other active participants in the systems design process, will incorporate concepts of human and organizational needs which are related principally to the organizational reality which he perceives and not to the view of reality held by employees and managers in the situation where computer systems are introduced.

If this hypothesis proved to be correct it would lead to a third hypothesis that these conflicting interpretations of reality will produce a bad fit between the design outcomes of computer systems and the efficiency and job satisfaction needs of the workers and managers who use them. The result, in the words of Stafford Beer [1972] can be a *non-viable* system. A system which cannot survive or which will only survive in a state of instability. Beer argues that "whatever makes a system survival noteworthy is necessary to it". It would seem probable that a necessary component of the successful creation and survival of computer systems which involve human beings is a definition of environmental reality by systems designers which equates reasonably well with the definition held by systems users – in other words which incorporates the same *models of man*. This similarity of perception, if it can be achieved, should produce a better fit between the efficiency needs of the enterprise, which must be met if it is to survive, and the human needs of those workers and managers who use the computer system. These too, must be met if they are to achieve psychological health and *job satisfaction*.

It was hoped that as a result of testing these hypotheses through the collection of data in a number of case study situations it would be possible to develop some insights into the role of values in the systems design process.

1.2. THE RISE OF THE RATIONAL ETHIC

In order to understand the present we need to know the past. This is especially so in a book on "values, technology and work" for all industrial countries appear to have been in the grip of a powerful ideology which is only now beginning to lose credibility. This ideology has seen labour as an expendable, easily replaceable commodity which produces at highest efficiency and lowest cost when few demands are made of it, when work is tightly controlled and when little or no discretion is allowed to the individual [Mumford 1978]. Although feudal society was built on what today we would see as class privilege and class oppression, these relationships were legitimized by a religious doctrine that taught that men were born into particular stations in life, and softened by the acceptance by the rich of a set of personalized obligations to the poor. Because commercial dealings were

4

based on personal contact the creation of trust and esteem was seen as important and the individual as a person could not be separated from his economic activities [ibid: 157]. Economic impulses were therefore tempered by diffuse social obligations which were a feature of medieval society.

Large scale industrial organization was preceded and influenced by the progressive secularization of the religious quest for truth which Carroll sees as the principal influence on thinking in Europe over the last three centuries [1974: 5]. In his view the desire for human redemption was recast so that the traditional routes for spiritual pilgrimage were transformed into a single path leading to the goal of rationality, or the dominance of order and control. Rationalism as a desirable objective therefore stemmed from many sources. It was a product of monastic asceticism, Calvinism and also of Newtonian physics. Rationalism, the drive to classify and quantify, was later joined by Utilitarianism, the drive to maximise material happiness, by the new economists, in particular Jeremy Bentham. It was Adam Smith who founded economics as a science, publishing the *Wealth of Nations* in 1775 and providing the example which inspired Bentham to extrapolate Newtonian methods into other spheres of social investigation [ibid: 6].

Bentham wanted to establish morals as an exact science and subject it to the rule of reason. Whereas the later work of Frederick Taylor was based on "l'homme boeuf", the strong man who could be assisted to use his strength to maximise his material prosperity, Bentham was influenced by "l'homme machine", a man reducible through scientific investigation to determined conditions. In this way "Bentham created the inspiration for economic planning, for 'social engineering' as we know it today" [ibid]. Bentham saw the individual as being wholly egoistic, seeking pleasure and avoiding pain. Morality was the act of being happy and the principle of utility was,

> that principle which approves or disapproves of every action, whatsoever, according to the tendency which it appears to have to augment or diminish the happiness of the party whose interest is in question [Bentham 1907].

For Bentham wealth was good because so much wealth produced an equivalent amount of happiness; the individual should therefore strive to acquire wealth. Similarly, the happiness of a community was a product of economic sufficiency for all its members. If the interests of society ever came into conflict with the interests of the individual than the two sets of interests must be brought into harmony through rational laws. Legislation was Bentham's instrument for turning society into a predictable, well-ordered economic system [Carroll 1974: 8].

Rationalism as a doctrine was therefore made explicit in the writings of eighteenth century philosophers and economists. It was based on a belief in human development and progress in which greater knowledge could lead to greater happiness. These ideas were given practical form by the engineering innovators of the time. In 1787, two years before the publication of Bentham's book *An Introduction to the Principles of Morals and Legislation*, James Watt produces his steam engine. Carroll says,

5

as the first man to apply technological principles concerning heat and mechanical energy to large scale work problems, he effectively bridged the gap between Netwonian models and man's practical struggle to control his environment [1974: 8].

The techological developments of the time therefore reinforced and gave practical form to the new philosophical ideas of economic efficiency and utility. Carroll suggests that motives other than economic also influenced the belief in rationalization. He saw these as psychological, a desire to establish a balance between order and chaos. The influence of Calvinism may have played a part here with its emphasis on discipline, morality and the virtue of work [Smith 1975]. In England Calvinism became Methodism and Thompson suggests that Methodism and Utilitarianism together, made up the dominant ideology of the industrial revolution [1968: 441]. In 1859 this philosophy was set out in Samual Smiles's *Self-Help*, a treatise on how the virtues of perseverance, energy, thoroughness and self-reliance will lead to economic and status rewards. These ideas also fitted the liberal philosophy of John Stuart Mill that, in addition to science and progress, there must also be a confidence in the resources of the individual [Carroll 1974: 13].

During the nineteenth century philosophy and technology continued to reinforce each other and in Charles Babbage we have an English technologist and mathematician who not only invented the computer, but also anticipated Taylor in his belief in the division of labour. In 1835 he wrote:

> *Skill acquired by frequent repetition of the same processes.*
> The constant repetition of the same process necessarily produces in the workman a degree of excellence and rapidity in his particular department, which is never possessed by a person who is obliged to execute many different processes. This rapidity is still further increased from the circumstances that most of the operations in factories, where the division of labour is carried to a considerable extent, are paid for as piece work . . . [1835: 119].

But these ideas did not go unchallenged. Carroll describes the group he calls anarcho-psychologists, led by such thinkers as Stirner, Dostoyevsky and Nietzche, who by the middle of the nineteenth century were starting the revolt against positivism. They saw the rationalist approach as leading to false goals and preferred means to ends, favouring creative activities such as adventure, variety and play [1974: 172]. At the same time Karl Marx was writing of the exploitation of the labourer in the factory and of how he was deprived of culture and of the opportunity of becoming a complete man [Struik 1965]. Marx, like Bentham, believed in a rational social order, but one of different kind from that he saw developing in Britain.

An influential critic at the end of the nineteenth century was Max Weber. Although younger than Marx – he was twenty when Marx died in 1883 – his descriptions of the bureaucratic patterns which result from the application of rationalization are similar to Marx's description of alienation under capitalism.

But whereas Marx saw alienation as a result of property relations, and the abolition of private property as the cure, Weber was more interested in the relationship of religion to the world that was being created by science and bureaucracy [Lystad 1972] and saw rationalization as an integral part of any industrial society whether capitalist or socialist [Fox 1974: 229].

As industry developed so economics became separated from ethics and utility from morality, and employment relationships became increasingly contractual [Fox 1974: 161]. During the later part of the second half of the nineteenth century the process of rationalization and mechanization began to speed up and this period saw the beginning of mass production industry and the fragmentation and deskilling of work previously performed by craftsmen. It also saw the appearance, in the United States, of the Scientific Management movement. Fox identifies two analogies in the literature of the time; the depicting of the human body as a machine, which can be made to respond like a machine, and the factory as a military establishment requiring both hierarchy and discipline [ibid: 192]. Taylor and his fried Gilbreth based scientific management on the earlier ideas of Babbage but, whereas Babbage had been a professor of mathematics at Cambridge, Taylor and Gilbreth were a practising engineer and a building contractor, respectively, and therefore in an excellent position to apply their ideas [Taylor 1947: 39].

Taylor's intention was very much in accord with the earlier philosophy of Bentham. Through applying intelligence and scientific method to what he saw as the muddle of inefficiency of industry he proposed to increase the profitability of the enterprise and the earnings of the employee. In this way both would gain and conflict would change to harmony. This improvement was to be achieved by separating "doing" from "thinking" and allocating the first to workers and the second to management. The role of management was to study scientifically the best methods for carrying out tasks and for measuring the level of performance reached. The role of the worker was to obey.

Davis has described the values and beliefs associated with the scientific management movement [1971]. First there was the belief that the human being could be treated as an operating unit which could be adjusted by training and incentives to meet the needs of the organization. Second, there was the view that people were unreliable, with narrow capabilities and limited usefulness; and this justified their being given small, low discretion jobs. Third, labour was a commodity to be bought and sold by the organization and fourth, a materialistic ethic suggested that if the end of increased material comfort was achieved then this justified the means to achieve it. These values are still to be found in much of British industry often supported by a trade union value system which regards any kind of work as acceptable providing it is safe, healthy and well paid.

Like the nineteenth century, the twentieth century has had its critics of the rational ethic. These are too many to mention here but range from psychologists such as Maslow, Herzberg, McGregor and Likert who believe that man can develop his potential within an industrial society, providing this is organized in a humanistic way, to Fromm and Freire who state that the oppressed must be helped to recognise their oppression so that they can rebel against it [Freier 1972: 21]. In between come the technology-watchers who focus on right and wrong uses of technology. These range from Sigmund Freud who suggested that the benefits of technology were "cheap pleasures" and "if there were no railway to make light of distances my child would never have left home and I should not need the telephone to hear his voice" [1953: 46] to Lewis Mumford who speaks of our "technocratic prison" [1964: 435]. This group sees technology as potentially anti-human. Slater in his book *The Pursuit of Loneliness*, a description of present-day American society, says: "The old culture American needs to reconsider his commitment to technological 'progress'. If he fails to kick the habit he may retain his culture and lose his life" [1971: 132]. Tom Lehrer gives his view of the technical ethic in song. " 'Once they are up who cares where they come down: thats not my department', says Werner Von Braun" [ibid: 130].

This book seeks to establish the extent to which a rational/technical ethic influences the design of computer systems in four organizations.

2. What are Values?

This book is about the values of individuals, groups and organizations and how these values affect both people's needs and expectations in the work situation and the way in which events there are interpreted. It is therefore important and necessary to describe clearly what is meant by the concept ''values'' when used in this kind of context. But, on beginning this task of definition the researcher immediately discovers that there is a superfluity of information, for the nature and role of values has been a subject of interest and discussion since the first Greek philosophers. The meaning of the term has been continually redefined from the years BC to the present day and has varied from a set of universally true moral precepts to a contingency approach in which values are adjusted rapidly and continually as new pressures in a culture alter people's views on what is desirable and undesirable [Bourke 1970: 105]. In order to cope with this information overload there has to be a selection of sources and although the ideas of some philosophers will be discussed in this chapter the views which will guide the analysis of values in subsequent chapters will be those of social scientists, particularly social scientists interested in theories of organization.

2.1. THE ORIGIN OF VALUES

In order to understand why we have values, why we approve of and want some things and not others we have to consider ideas from philosophy and psychology. Philosophers have already recognised and separated two kinds of problem. There are first the problems of how things are, what is a person and what sort of world this is. These are the problems of ontology. Second are the problems of how we know what sort of a world it is and what sort of creatures we are that can know something of this matter. These are the problems of epistemology [G. Bateson 1972: 284]. Ontology and epistemology are inextricably bound together, for a person's beliefs about what sort of a world he lives in will determine how he sees this world and behaves in it. Similarly, the way he perceives and behaves will determine his beliefs and interpretation of the world. This interaction between beliefs and vision and interpretation and behaviour has been called ''cognitive structure'' or ''character structure'' [ibid: 285]. This term expresses, although inadequately, the fact that people constantly make assumptions and build premises about the world in which they live and their own relationship to this world. These assumptions and premises may be objectively correct, assuming

9

there is such a state as objectivity, or they may be incorrect and based on what "the man on the Clapham omnibus" would regard as erroneous visions and interpretations. However, whether true or false they tend to be self-validating, for, if a person sees a situation in a particular way, this perception will determine his beliefs about the situation and his behaviour in relation to it. People who inhabit a particular culture or possess a certain level of awareness may interpret the world in a totally different way from those who live in other cultures or attain other levels of awareness. A good example of the first can be found in the writing of the American anthropologist Carlos Castaneda who is taught by the Yaqui Indian magician, don Juan, to "see" the Indian world; something that Castaneda finds extremely difficult even when assisted with drugs. For example, after his first experience of smoking peyote Castaneda says to don Juan:

> "I really felt I had lost my body, don Juan"
> "You did."
> "You mean, I really didn't have a body?"
> "Well, I don't know. All I can tell you is what I felt"
> "That's all there is in reality – what you felt."

This conversation continues for some time and Castaneda later writes in his research notes:

> It was useless to persist in trying to get a rational explanation. I told him I did not want to argue or to ask stupid questions, but if I accepted the idea that it was possible to lose my body I would lose all my rationality. [Silverman 1975: 4]

An example of the second is provided by Timothy Leary, a psychologist who has experimented with psychedelic drugs. He tells us:

> Ontologically there are an infinite number of realities, each one defined by the particular space-time dimension which you use. From the standpoint of one reality, we may think that the other realities are hallucinatory, or psychotic, or far out, or mysterious, but that is just because we're caught at the level of one space-time perception. [Leary 1970: 195]

Kelly, a psychologist, regards the assumptions and premises associated with "seeing" and understanding as the rules by which an individual construes or interprets his experiences; "each man contemplating in his own personal way the stream of events upon which he finds himself so swiftly borne" [Kelly 1963: 3]. This is not a new idea but one that is found in the work of many writers, psychologists and philosophers such as William James, Freud and John Locke who, we are told, sat down one night to write "An essay concerning human understanding" and finished it twenty years later. Psychologists such as Kelly who are interested in what are called "personal constructs", the way we observe, interpret and make sense of the world in which we live, suggest that we do this through a process of what today cyberneticians would call "variety reduction or control" [Beer 1966: 316]. That is, we create in our own minds a series of simple

10

models or patterns (Kelly calls these transparent templets) of expected relations between things, and we continually compare events in the real world against these conceptual models. If the fit between the mental model and the real world situation seems a good one then this reinforces the validity of the model in our own minds and we continue to use it. If we find that reality consistently deviates from the model then we adjust the model to ensure a better fit. These models or patterns have been described as

> ways of construing the world. They are what enables man, and lower animals too, to chart a course of behaviour, explicitly formulated, or implicitly acted out, verbally expressed or utterly inarticulate, consistent with other courses of behaviour or inconsistent with them, intellectually reasoned or vegetatively sensed. [Kelly 1963: 9]

A good fit between one's mental models and the real world, or between what one expects and what does happen, provides intellectual security. Events become understandable and predictable and one can estimate the consequences of particular courses of action. Because this intellectual security is felt as a desirable state, for it produces feelings of knowledge and control over events, man constantly tries to increase his repertory of models so that he can understand an increasing range of situations. He also tries to refine these models so that they provide ever more exact explanations of his personal world and give him even better guides to action. People who are extremely good at adjusting their models of the world to new experiences we regard as quick-witted, responsive and intelligent. Those who hold on to a set of old models when they no longer provide good explanations for real world events, we see as narrow, conservative or even stupid.

As a rule individuals do not keep these mental models of the world to themselves, they check them out by discussing them and establishing that others are seeing the world and interpreting events in a manner similar to themselves. Through this communication process, which we call language, these constructs become widely shared and people within a particular culture, who share the same experiences, come to interpret reality in similar ways. Outsiders, however, may have great difficulty in comprehending the reality of groups other than their own. Castaneda makes this point when commenting on his anthropological research data.

> To any beginner, Indian or non-Indian, the knowledge of sorcery was rendered incomprehensible by the outlandish characteristics of the phenomena he experienced. Personally, as a western man, I found these characteristics to bizarre that it was virtually impossible to explain them in terms of my own everyday life, and I was forced to the conclusion that any attempt to classify my field data in my own terms would be futile.
>
> Thus it became obvious to me that don Juan's knowledge had to be examined in terms of how he himself understood it; only in such terms could it be made evident and convincing. [Castaneda 1968: 19]

People in groups not only come to interpret events in similar ways, they identify certain states, situations or outcomes as desirable and others as undesirable. There is then a move from using mental models purely in terms of explanation or prediction and towards associating choice or judgement with these outcomes. "We prefer this to that"; "we want to achieve this, rather than that". In this way values become part of mental models.

Some mental models do not fit well with present reality, yet are powerful, enduring and difficult to change, and this kind of model often has a strong value element contained within it. Factory workers may designate management "our enemy", even though the behaviour of their management is consistently benevolent. This kind of model incorporates a powerful value judgement often based on traumatic past experiences, which more recent experiences have had little influence in changing. However, what is called the "theory of constructive alternativism" suggests that all our present interpretations of the world are subject to revision or replacement [Kelly 1963: 15], even though some are revised or replaced more rapidly and more easily than others. Kelly tells us:

> Man comes to understand his world through an infinite series of successive approximations. Since man is always faced with constructive alternatives, which he may explore if he wishes, he need not continue indefinitely to be the absolute victim of either his past history or his present circumstances . . . . Constructs are tested out by the individual in terms of their usefulness in helping to anticipate the course of events which make up the universe. The results of the testing of constructs determine the desirability of their temporary retention, their revision, or their immediate replacement. [ibid: 44]

If we accept that values stem out from our mental models or constructs then an acceptance of Kelly's argument would lead us to the position that all values have the potential to be changed and that there are no fundamental values that endure for ever. Most social scientists would accept this argument although some philosophers and many priests would not.

Our argument here is that values emerge from the mental models which people use to understand and make sense of their personal worlds. These models are revised when experience consistently deviates from the model so that it loses its value as an aid to prediction. As these models are communicated from one member of a group to another, for example, from parent to child, and their use as explanatory aids and guides to action discussed and affirmed, or asserted and internalised, so sets of common constructs emerge and are accepted by the group as a whole. The need to achieve particular goals or objectives leads a group into value judgements of what is "right or wrong", "desirable or undesirable" and these value concepts are then incorporated into the mental models as guides to choice and action. General agreement on values is thought to cement a group together and in this way to assist its survival [Dahrendorf 1959: 157].

2.2. A DEFINITION OF VALUES

Because of the imprecise nature of the notion of values a wide range of definitions has accompanied the use of the term. These extend from the very general "value is any object of any interest" [Perry 1926: 115] to precise definitions associated with beliefs in absolute moral values in which value is conceived as "something wholly independent of man's feelings and identified with certain objects that are quite distinct from the ways in which we react emotionally to them" [Schlick 1939: chap. 5], for example, fidelity, gratitude, non-malevolence [Ross 1930: 3]. Most definitions fall broadly into one of two categories: the first defines value as "something" – a property, a relation that an object has, as in economic theories of value. The second sees the attribution of value as an activity that occurs in someone's mind [Lamont 1955: 3]. This mental process is subjective and related to making choices and it is often called a theory of valuation in preference to a theory of value. In this book we are concerned with the second category of definition and interpret the word values as relating to ideas which make us consider given objectives, qualities or events as valuable [Najder 1975: 42]. Values when applied to ideas, principles, or criterions enable us to evaluate particular occurrences and to ascribe to them a positive or negative value. In this way our values help us to make judgements [Arrow 1951: 18]. For the purpose of this study we like and propose to use the definition of value provided by Kluckhohn, *"a conception explicit or implicit, distinctive of an individual or characteristic of a group, of the desirable which influences the selection from available modes, means and ends of action"* [Kluckhohn 1951: 395].

Kluckhohn is careful to define his terms using the word *conception* to indicate that "value" is a logical construct comparable to culture or social structure. Values, like culture, are not directly observable but are based on the interpretations people make of their personal worlds and how these interpretations are translated into word and action. The definition contains the words *explicit* or *implicit* to indicate that some of the deepest and most pervasive values may be only partially and occasionally communicated through language. They have become internalized and built into the fabric of the culture. The word *desirable* tells us that value statements are normative and are related to things which may not exist but which people think should exist. Values are always associated with a degree of commitment; a thing is approved of and therefore it is wanted and striven for. The word *selection* is used in preference to choice as Kluckhohn regards it as more neutral. Selection or choice is always restricted by a limitation of resources of which knowledge may be one [Mumford and Pettigrew 1975: 39].

The fact that choice is associated with values provides the opportunity for research. It is possible for a researcher to identify many of the things a group regards as desirable, to establish the nature of the different courses of action available to it to achieve these objectives, the kind of choices group members make and the success of these choices in achieving the objectives. This kind of evaluation will be applied to a number of case study situations in this book.

Values, like other aspects of consciousness, require evaluative systems; mental models that assist ways of perceiving and of reasoning; that help the interpretation of personal experience and provide a basis for making choices. Philosophers in particular are interested in the kinds of mental constructs that produce the choices we make. They wish to understand the nature of the activity which they have termed "evaluation" [Lamont 1955: 4], what mental processes lie behind the awarding of approval and why some things are approved of and designated "good" while others are not seen in this way?[1] To answer these questions it is helpful to consider again Kluckhohn's definition of values, this time attaching to it an additional sentence which brings in the notion of different mental processes. Kluckhohn defines values as:

> a conception explicit or implicit, distinctive of any individual or characteristic of a group, of the desirable which influences the selection from available modes, means and ends of action . . . .

He continues:

> It must be emphasised, however, that affective (desirable), cognitive (conception) and conative (selection) elements are all essential to this notion of value. [Kluckhohn 1951: 395]

Kluckhohn here uses an old terminology which distinguishes three aspects of mental life as cognitive (knowing, etc.), affective (feeling) and conative (willing, wanting and suchlike) [Lamont 1955: 212].2 These in turn appear to be derived from theories of the tripartite soul developed by the early Greek philosophers in which two levels of the psychic soul are distinguished: the rational and irrational [Plato, *The Republic,* Book IV]. The rational is that part which reflects and acquires knowledge. In the irrational are two parts: one feels anger, indignation, and the ambition to excel – it is the spirited element; the other part desires the pleasures of food, sex and so on – it is the element of desire or appetite.

Philosophers would accept Kluckhohn's proposition that affective, cognitive and conative elements are all associated with the choices and approval implicit in value judgements. But they are interested in which of these elements predominate. Some ideas on this issue will also be useful to an understanding of the values of the groups in our research situations and so the arguments are set out briefly here. Lamont provides a useful summary of these [1955: 209]. He begins by asking the question whether approval, which is essentially a subjective activity, can be regarded as cognitive and based on a rational evaluation of a

---

1. In case readers wish to challenge the use of the word "good", it is used here in the sense offered by Ross who says, "The universal precondition – and apparently the only one – of our using the word 'Good' is the existence of a favourable attitude, an attitude of commendation or approval in us towards the things in question" [Ross 1930: 140].

2. Conative is being interpreted in somewhat different ways here. Kluckhohn defines it as "selection", Lamont as "willing, "wanting", etc. Both of these definitions imply action however, and to "will" you must first chose what you want. In Latin the word *conatus* means "effort", "endeavour".

choice situation. His conclusion is that to the extent that a judgement of goodness or badness, desirability or undesirability is involved, this is a cognitive response, but he considers that value choices are more than this. The fact that something is approved of implies an active tendency to bring it into existence or to maintain it in existence if it is already there. This is a conative disposition for it contains a dynamic of creating, maintaining or destroying. Therefore in Lamont's view conative elements are stronger than cognitive elements in value judgements. However, some philosophers have proposed that approval is an affective attitude; that when we attribute goodness to things we are expressing our feelings or emotions [ibid: 213]. The argument against this is that judgements are not based on "feelings"; rather "feelings" are consequent upon judgements being made; therefore the cognitive element in a value judgement must be stronger than the affective element [Ross 1930: 131]. Lamont's conclusion is that approval is essentially a conative attitude, although it contains strong elements of cognition. He argues:

> When we speak of cognition we are thinking of those aspects of physical life, such as perceiving, reasoning or believing which are concerned with the apprehension of things as they are, were or will be; and when we speak of conation, we are thinking of those aspects such as willing, striving and desiring, which operate as "forces" determining what the nature of things will be. Cognition is "informative about" the objective order; conation is actually or potentially "formative of" the objective order. [Lamont 1951: 228].

Lamont accepts with Kluckhohn that approval, and therefore values, usually includes all three elements; that we are concerned with a total state of mind which simultaneously contains processes of "knowing", "feeling" and "willing" (or "choosing"). Lamont's view is that in our inner experience we draw significant distinctions between "knowing", "feeling", "willing" and therefore it is useful to know which aspect is being emphasized when we are said to "approve" rather than to "know", "will" or "feel" [ibid: 238]. He chooses conation because it is "that attitude or aspect of our mental state which is practically *activist*; it is not 'informative about', which would imply a cognitive emphasis; but actually or potentially 'formative of' because it seeks to maintain or create an approved state of affairs". Value judgements imply approval and approval implies an action state of creating or maintaining.

To the social scientist interested in evidence as much as argument these logical propositions may appear to be erected on flimsy foundations. Nevertheless the propositions put forward by philosophers do throw light on the nature of values. Also the ideas of cognition, conation and affectivity have been used as analytical tools by Talcott Parsons [1951: 75] and will be used in this way in this book.

2.3. VALUES AND THE SOCIAL SCIENTIST

It is clear from our discussion so far that the notion of "values" is complex and diffuse. Those who have written on the subject have done so from a variety of different intellectual positions and a great deal of this writing appears to have little relevance to the needs of social scientists [Meehan 1969: 2]. Social scientists therefore have to create an approach to value questions that is relevant to their interests, particularly those related to "the kinds of problems that man must somehow solve in order to live in society with other men" [Meehan 1969: 2]. Many social scientists have contributed descriptively to our knowledge of the values of individuals, groups, organizations and societies but the work of Kluckhohn and Parsons stands out in that it makes a major contribution to a theory of values that has relevance to the social sciences. Because "man is an evaluating animal" [1951: 403], it would be impossible to provide an adequate description of an individual or a group without a statement about the individual or the group's values [Najder 1975: 149]. Values are of concern to sociologists, psychologists, anthropologists, economists, historians and political scientists; they can therefore act as an integrating concept within the social sciences. Kluckhohn quotes a discussion of a Value Study Group at Cornell University, as follows:

> The concept "value" supplies a point of convergence for the various specialised social sciences, and is a key concept for the integration with studies in the humanities. Value is potentially a bridging concept which can link together many diverse specialized studies – from the experimental psychology of perception to the analysis of political ideologies, from budget studies in economics to aesthetic theory and philosophy of language, from literature to race riots . . . . [ibid: 389].

Value judgements can only be understood within a specific empirical context [Meehan 1969: 45] and for the sociologist this context will be culture and group. If we return to Kluckhohn's definition of values we find that this is made apparent, for his complete paragraph reads:

> A value is a conception, explicit or implicit, distinctive of an individual or characteristic of a group, of the desirable which influences the selection from available modes, means and ends of action . . . . It should be emphasized here, however, that affective (desirable), cognitive (conception), and conative (selection) elements are all essential to this notion of value. *This definition takes culture, group, and the individual's relation to culture and place in his group as primary points of departure.* [1951: 395]

Bateson contributes a cultural element to our earlier discussion of mental models or personal constructs when he comments:

> The individual is needlessly simplifying, organizing and generalizing his own view of his own environment; he constantly imposes on this environment his own constructions and meanings; these constructions and meanings are characteristic of one culture, as over against another. [Hunt 1944: 723]

Both the cybernetician and the psychologist might argue with Bateson's use of the word "needlessly", for they would regard this process as essential to what the first would call "homeostasis" and the second "mental equilibrium" or "sanity".

16

The "team" or the community, or even the wider society, is defined by a common set of values and categories. We recognise groups of people as communities or societies because they share ideas about how things are and how they should be: they have a common set of categories with which they describe the social and natural world around them and an agreed definition of the good things and the bad things in life [Bailey 1971: 8]. Kluckhohn uses the term "value orientation" for value notions which are (a) general, (b) organized and (c) include existential judgements, by which he means judgements relating to the real world. A value orientation is a set of linked propositions which embraces both values and judgements about action. He defines this as:

> A generalized and organized conception, influencing behaviour, of nature, of man's place in it, of man's relation to man, and of the desirable and non-desirable as they may relate to man-environment and interhuman relations. Such value orientations may be held by individuals or, in the abstract-typical form, by groups. Like values they vary on the continuum from the explicit to the implicit. [Kluckhohn 1951: 411].

2.4. CLASSIFYING VALUES

Classifications are a help both to comprehension and research. They identify distinctions and diversity and assist an understanding of relationships, complexity, and similarities and differences. It is suggested that one cannot begin a really well-informed discussion of any range of phenomena (dogs, games, diseases, etc.) unless at least some rough classification is at hand [Rescher 1969].

The most useful classification is one that relates clearly to the nature of the problem that is being studied and no classification should be regarded as fixed. If a classification is used to assist the definition of a research problem then a useful output of the research may be an improved classification as greater insight is achieved. Rescher has provided a number of suggestions on how values may be classified. He suggests [ibid]:

Classification by who subscribes to the value

Is the value held or should it be held, by an individual or a group; and what sort of individual or group?

Classification by reference to what the value is applied
For example,

Name of value type	Explanation of what is at issue	Sample values
Thing values	desirable features of inert things	speed of car
Environmental values	desirable features of the environment	landscape, beauty
Individual or personal values	desirable features (character traits, abilities, talents)	bravery, intelligence
Group values	desirable features of the relationship between an individual and his group	respect, mutual trust
Societal values	desirable features of arrangements in society	equality

Classification by the nature of the benefit associated with the value

Category of value	Sample value
Material and physical	health, comfort, security
Economic	economic security
Moral	honesty, fairness
Social	charitableness, courtesy
Political	freedom, justice
Aesthetic	beauty
Religious	piety
Intellectual	intelligence
Professional	professional recognition
Sentimental	love

Classification by the purpose of the value

    "value for *bargaining* purposes"
        persuasive
        deterrent

There are many other possible ways of classifying values; for example, by the relationship between the holder of the value and the perceived beneficiary, as with parent-child values. Or by the relationship the value itself bears to other values, in which case its characteristics would be described in detail and compared with the characteristics of other values.

A classification that resembles this last category and which fits well with the interests of social scientists has been developed by Kluckhohn [1951: 413]. This analyzes values in terms of their "dimensions", using the term in its mathematical sense of a phenomenon that can be measured. These categories are as follows:

Dimension of modality

Positive or negative values related to attitudes of approval or disapproval.

Dimension of content

Aesthetic, cognitive and moral values. Aesthetic is related to the term "affective" in Kluckhohn's definition of values; it refers to feelings and emotion, particularly those associated with beauty. Cognitive we have already defined as values which are a product of rational judgements. The introduction of "moral" values here seems to be based on an assumption that in "selecting" or "wanting" – the cognitive element – we strive to realise what we believe to be "good". This does not imply that we are responding to any formalized system of ethics as goodness can be defined simply as those things of which we approve [Ross 1930: 254].

Dimension of interest

These are the values associated with the best way or *mode* of doing something. They are sometimes called "expressive" values as they are concerned with means rather than ends. We express a disposition to achieve something [Lamont 1955: 249].

Dimension of generality

Some values are specific to certain situations or things, others have a wide range of application.

Dimension of intensity

Kluckhohn believes that the strength of a value may be determined by observing the amount of energy expended in maintaining or achieving it and also the fierceness of the sanctions imposed for any contravention. He suggests that the method of paired comparisons is particularly helpful in measuring the strengths of values [Kluckhohn: 414].

Another useful notion is that of *central* and *peripheral* values which differ according to the number and variety of behaviours influenced and the extent to which a group or individual would be markedly different if the value disappeared. However, when estimating the intensity of values and conformity to them one must be careful not to confuse variation with deviation. Most cultures accept a range of conformity for people in different age, sex and class groups, and only when behaviour moves outside these accepted limits is it regarded as deviant and subject to sanctions. Values include *prescriptions, permissions* and *prohibitions* (what we must do, what we may do, what we must not do).

Dimension of explicitness

In general an explicit value is one that is communicated verbally and is widely and openly recognized. An implicit value is hidden, assumed and internalised. Implicit values may be difficult for outsiders to recognize.

Dimension of extent

An idiosyncratic value is held by one person only in a group. Idiosyncratic values are important because their existence is one way in which new values come into being and these may be necessary to deal with unexpected situations and changing environments. In contrast, group values are normative and usually derived from a group's history and social structure.

Dimensions of organization

This is the question of whether values are organized hierarchically. Can values be placed in some kind of pyramid structure? Certainly, in most situations some values will be seen as more important than others and as a research exercise people can be asked to rank their values. It may not, however, be very enlightening if they do so. Kluckhohn argues that what matters is not so much the hierarchy as the logical and meaningful relation of one value to another. *General* values, or values which are powerful and universally accepted, are likely to have both high priority and many relationships with other values; in other words a quality of "connectedness" [Podger 1978]. It may be useful to distinguish *isolated* values which neither conflict with nor support other values, and *integrated* values which form part of an interlocking network.

Kluckhohn concludes his classification by pointing out that his categories have been arrived at theoretically. Yet values can only be understood in terms of particular events and detailed descriptions of these events. Such a description will be found in the case study chapters of this book.

2.5. CHOOSING VALUES

We have seen that values are, and can be, defined in many different ways. The only area of general agreement is that values are "normative", they are concerned with what should be rather than what is. Values therefore concern consequences and values have to be chosen which produce the kinds of consequences that an individual or group desires. This idea comes from Aristotelian ethics which are concerned with the well being (*eudaimonia*) and are teleological in that they stress the purposiveness of human nature [Bourke 1968: 33]. Before a person takes action he will weigh up the situation he is involved in and then apply a value judgement to guide his action. The situation is defined cognitively but it is interpreted affectively by the individual through a consideration of the kind of action that will further his personal well-being. The selection of a course of action and the "will" to embark on it come next as a conative response.

Choosing between values may not be an easy activity, particularly if the choice is not between things which are "better" or "worse" but among a number of values each of which leads to a different kind of well-being. Najder points out:

> To choose between values one has to be aware of a possibility of choice and have the practical ability to reach a conscious decision . . . [one requires] an ability to compare the values from among which a choice is feasible. [1975: 158]

Decision theory tells us something of how these choices are made [Mumford and Pettigrew 1975: chap. 3].

Many value systems are "consequentialist" in that the prime concern of the individual or group is the practical consequences of any action taken and these are carefully evaluated. Other, more traditional, value systems are based on rules and social duties which are binding irrespective of the consequences. Values are clearly evolutionary, for men constantly test out and modify their values according to the results obtained from their use [Meehan 1969: 39] and some commentators argue that we are increasingly moving away from non-consequentialist and towards consequentialist value systems. Gowler, for example, argues that given the increasingly temporary nature of social rights and duties and the related fragmentation of social life, behaviour, and consequently values, are becoming more and more contingent. Values are ceasing to be normative and becoming relative. What is "right" or "wrong", desired or not desired becomes increasingly less absolute and more a response to the pressures exerted by a constantly changing environment [Ruddock 1973]. Nevertheless, whether we choose our values pragmatically on the basis of a careful evaluation of where they will take us, or accept them unquestionably because they are a part of our social fabric, motivationally they attract us by the visions of the ideal situations they evoke and by our belief that the acceptance of a given value will produce desirable situations or states of consciousness [Black: 145].

Opinions differ on whether the number of choices available to us are increasing or decreasing. Jacques Ellul, a French philosopher, claims that man was freer in the past when "choice was a real possibility for him . . . . In the future, man will apparently be confined to the role of a recording device": devoid of choice he will be acted upon, not active. He will live in a totalitarian state run by a velvet-gloved gestapo [1967: 77]. Alvin Toffler, by contrast, suggests that the problem of the future will be one of over-choice, for a more sophisticated technology produces greater, not less variety. Like Gowler he sees the values of the future as being "shortlived and more ephemeral than the values of the past". He believes that "there is no evidence that the value systems of the techno-societies are likely to return to a 'steady state' condition". For the foreseeable future, "we must anticipate still more rapid value change . . . we are witnessing the crack up of the consensus" [1971: 277]. Toffler in common with other commentators, believes that the increased complexity of modern life is likely to produce either confused values or an absence of values. He argues, "presented with numerous alternatives, an individual chooses the one most compatible with his values. As over-choice deepens, the person who lacks a clear grasp of his own values is progressively crippled" [ibid: 376].

Pitrim Sorokin in a foreword to Maslow's book *New Knowledge in Human Values* is even more pessimistic believing that we are moving towards a form of society without values. He says, " . . . the ultimate disease of our time is valuelessness . . . . The state of valuelessness has been variously described as anomie, anhedonia, rootlessness, emptiness, hopelessness, the lack of something to believe in and to be devoted to" [Maslow 1959].

Another view is put forward by the protagonists of alternative life styles who argue that we must all create our own values. For example, Timothy Leary suggests: "The standard operating procedure for a life of ecstatic prayer and exultant gratitude is to write your own Declaration of Independence and constitute your own vision of the holy life" [1970: 300]. Even more forcibly he tells us, "Start your own religion (Sorry, baby, no one else can do it for you)".

The choice aspect of values therefore seems inextricably linked to the structure of the society in which the values occur. Stable, slow moving societies have values which are internalized and enduring and in which choice is not a conscious process. Present-day industrialized societies present their members with a wide variety of alternatives and values are related to consciously desired outcomes. Two problems now occur. The first that of knowing the available choices, the second that of evaluating which value when implemented will lead to the desired result. Future society is difficult to predict and we are offered widely differing possibilities ranging from too little choice to too much choice, together with a plea for the right of the individual to create and hold onto his own values, whatever they may be.

Because "choice" is associated with values the social scientist is presented with an opportunity for research, for although values are abstract and subjective they manifest themselves through the ways in which people talk and behave [Rescher 1969: 24], and this can be objectively studied. The social scientist must be able to determine how a value is seen both by the individual holding the value, and by the group to which he belongs [Kluckhohn 1951: 402] for the two interpretations may differ. There is often a rivalry between the spontaneous definition of a situation made by an individual and the definition with which his group provides him. It is suggested that the individual tends to a hedonistic choice of activity, putting his own pleasure first, while the group prefers a utilitarian choice which emphasizes safety rather than pleasure [Thomas 1957: 41]. It is also important to study value choice from the point of view of a disinterested observer. Choice then becomes a process of selection from a range of possibilities, many of which may not be visible to either the individual or the group to which he belongs. These three perspectives may overlap or diverge in differing degrees [Kluckhohn 1951: 402].

2.6. CONCLUSIONS

In this chapter an attempt has been made to "set the scene" for the analysis of empirical research data that will follow in chapters 6 to 12. At the beginning of the chapter theories of personality were used to explain how values became part of our cognitive structures and ideas derived from philosophy to clarify the concept of "value". For the social scientist values can only be meaningful within a cultural context and so the ideas of the two social scientists who have contributed most to the theoretical basis of this book, Clyde Kluckhohn and Talcott Parsons were discussed. In the next chapter Parsons' pattern variables will be used as the basis of an analytical framework for examining the "fit" between organizational and employee values. When this fit is a good one it is perceived as leading to what can be regarded as a set of mutually beneficial relationships between employees and the organizations for which they work.

3. Theoretical Tools

3.1. A FRAMEWORK FOR THE STUDY OF VALUES

In the last chapter attention was directed at gaining an understanding of the concept of "values", and obtaining some insight into the nature and function of values. For the purpose of this study, however, it is necessary to go further than this and to develop an analytical framework within which it will be possible to study the values of a number of different groups, and to establish how values affect the attitudes and behaviour of group members. The contribution made by both Clyde Kluckhohn and Talcott Parsons to a theory of values which is of use to social scientists has already been commented on and in this section an attempt will be made to operationalize this theory so that it can be applied empirically in a number of research situations. The ideas of both Kluckhohn and Parsons, although derived from earlier philosophical thought, are well suited to an examination of values within a cultural context. Also important from the researcher's point of view, the ideas of the two men reinforce and complement each other. Kluckhohn's contributions tend to be at the meta system level in that they are concerned with defining and describing values, Parsons is concerned with values but also with the lower level "needs" and "expectations".

Values are not the same as needs, although needs appear to arise from values. For example, the value of individualism forms an important part of American and Western European culture. From this value stem needs for self-expression and for personal freedom, these in turn generate needs for choice and choice opportunities [Lee 1948: 391]. Needs then are at a lower level than values, although values can only be operationalized in the context of a need.

Parsons uses the term *need dispositions* for sets of needs that lead to positive action and therefore are associated with motivation and choice [1951: 9]. Parsons sees the need dispositions of an individual as being integrated into an "on-going" personality capable of some degree of self control and purposeful action [ibid: 19]. Another useful concept developed by Parsons is that of *role expectations*. Individuals and groups are expected to behave in particular situations in certain specific ways. If one person fulfils the role expectations of another he will receive rewards in the sense of favourable attitudes and responses. Non-fulfilment of role expectations will usually lead to sanctions of various kinds. People associating together in social groups will tend to have similar patterns of needs dispositions and role expectations, which stem from similar value orientations.

24

It will be remembered that a value orientation has been defined as values which are generally held, which are organized in the sense that they contain an integrated view of a particular aspect of the world and which also involve judgements about appropriate action in the real world (see p. 17).

One of Parsons' main objectives has been the development of a theory of action which is related to values and value orientations. He sees the actions of an individual, a group or a society as influenced by a number of conceptions. The first is what the individual, group or society wants from a particular situation; the second is how they understand the situation and the third, how they intend to use the situation to get what they want from it. In other words how they develop their plan of action. Parsons believes that we all evaluate situations in terms of two things.

1. what we expect to happen: this is a cognitive and intellectual response directed at understanding a situation before taking action;
2. what we can influence in the situation in order to give ourselves a choice of outcome: this is an evaluative response directed at perceiving action opportunities. The subsequent action then being taken in terms of some moral position [ibid: 11].

Parsons thus sees an individual in a situation as being presented with a number of choices which must be made before the situation becomes clear and meaningful and he can take action. Parsons also sees a group or a society as having *role expectations* of its members so that it expects them to choose in a way that is in line with the values and interests of the larger group. In this study the larger group with which we are concerned is the industrial or commercial organization and so we shall restrict our remarks to the *need dispositions* of employees and the *role requirements* of employing organizations.

The choices which an individual or organization makes can be categorised into five dichotomies which Parsons calls "pattern variables". Pattern variables are the Parsonian term for a tendency to choose one thing rather than another in a particular type of situation or, from the organizational viewpoint, to expect one's employees to choose one thing rather than another.

Parsons names his pattern variables:

Affectivity	— affective neutrality
Self-orientation	— collectivity orientation
Universalism	— particularism
Ascription	— achievement
Specificity	— diffuseness

[ibid: 77]

These pattern variables cover the following choices:

At the *individual or personality* level, affectivity is the need to take advantage of an opportunity which presents itself in a situation in order to secure some immediate gratification. *Affective neutrality* is a willingness to defer short term gratification in the interests of some perceived longer term gain.

25

At the *group or organizational* level, *affectivity* is the "role expectation" that an individual can go for an immediate gratification and does not need to resist this in the interest of group discipline. *Affective neutrality* is the role expectation that the individual will refrain from striving for personal gratification in the interest of group discipline. An *affective* object is therefore one which attracts an individual because he perceives that he will derive a gratification such as pleasure or enjoyment or some other form of emotional response from it.

At the *individual or personality* level, *self-orientation* is the need to pursue a particular personal interest or goal without any regard for wider group interests. A *collectivity-orientation* is the need to take into account shared group values which the individual feels he has an obligation to adhere to.

At the *group or organizational* level, *self-orientation* is the role expectation that the individual can pursue his own long-term goals and interests. A *collectivity-orientation* means that there is an expectation that the individual will take account of the values and interests of the group.

At the *individual or personality* level, *universalism* means that the individual has a need to respond to situations in conformity with some generally accepted standards, *particularism* means that he is less conformist and will weigh up situations himself and judge them on the basis of his own perceptions and standards.

At the *group or organizational* level, *universalism* means a role expectation that when the individual is faced with choices he will make these on the basis of generally held rules and standards. Similar choice situations will therefore always be treated in the same way using the same criteria for forming judgements. *Particularism* means that when choices are being made, for example, for promotion purposes, the criteria used will be related to the applicant's special attributes such as skills and knowledge and not to some blanket criteria such as length of service.

At the *individual or personality* level, *ascription* means that the individual has a need to respond to the essential nature of an object rather than to its past, present or future performance. Thus if the object is a human being he will be responded to because of his personal qualities such as knowledge, humanity, friendliness and not because he has had some outstanding achievements. *Achievement* is the opposite of this and means that the factor which most influences the individual's response is performance. A man will be admired because of what he has done not because of what he is.

At the *group or organizational* level, *ascription* means that the role expectation is that an individual will respond to others because of what they are rather than what they do, an *achievement* orientation requires the individual to give priority to actual or anticipated performance when he is making judgements.

At the *individual or personality* level, *specificity* is the need to always respond to a situation in the same specific way. If the situation in question is a job then the individual will prefer tightly structured work with clearly defined boundaries. *Diffuseness* is a need to respond to situations in terms of the requirements of those situations, taking into account the interests and obligations of the individual making the choice. This implies a degree of open-mindedness and an ability to analyze situations. If the situation is once again a job it implies a liking for a loosely

26

structured work role which requires a variety of judgements and decisions [ibid: 81].

These pattern variables have the potential to provide a logical framework for examining values and needs in our research situations. The fact that they are applicable to the needs of the individual and to the role expectations of the employing organizations means that while they are particularly appropriate for examining the fit between what the individual is seeking from work and what he is receiving, they can also be applied to the needs of the organization. Parsons points out that all relationships contain what he calls a *complementarity of expectations* so that as well as ascertaining the extent to which the behaviour of an organization is meeting the needs of its employees, it is also possible to establish the extent to which the behaviour of employees is meeting the needs of the organization; these needs being defined by top management.

In this study we are interested particularly in the impact of values on the design, implementation of, and response to, technology, and a small pilot study carried out by the author and Hedberg, a Swedish researcher, has already provided evidence that a powerful influence on the design of computer systems could be the values of the systems designers [Hedberg and Mumford 1975: 31]. These influence their view of the user and his skills and abilities and generate preconceived ideas on the best ways of organizing work so as to reduce the likelihood of the human being making errors or acting irrationally. In other words, when they design new systems of work, systems designers have in their minds some individual and organizational models of man which they take as their frame of reference. These models are not necessarily conscious structures, for systems designers may be unaware that they hold them.

If values are the dominant, or one of the most dominant factors influencing attitudes and behaviour within organizations then it seems necessary to find out more about these. What kinds of values do the different groups working there hold and how do these values influence the way in which organizations are structured and managers relate with employees and define the role of employees. Similarly, how do the values of employees affect their behaviour in work and their relationships with their employers. Are there any shared values which serve to integrate the individual with the organization and are there areas in which values are in conflict and the individual is alienated from the organization and its objectives; it serves merely to provide him with a pay packet and perhaps a milieu in which he can vent his aggression on management. It seems that a framework of analysis is required which, following Parsons, concentrates on values as determinants of action, feelings and beliefs. The necessity for careful identification and examination of values has been stressed by Gowler [1974: 4-15], who makes the point that change will only work if it fits with the "values, beliefs and norms" of people. An important problem for the sociologist who wishes to understand the processes of change is to establish how these values, norms and beliefs affect the way in which people respond to change and how they influence the thinking of management when it designs and implements organizational change.

3.2. A FRAMEWORK FOR IDENTIFYING AND MEASURING VALUES

The problem now is to find or develop an analytical model which assists an understanding of values and their influence on behaviour. Here it is useful to look closely at Parsons' pattern variables for these represent his view of those values which most influence an individual or a group when it has to make a choice on how to behave in a particular situation. We need to know what aspects of a person's behaviour are a product of his beliefs and feelings as a person and what aspects are a result of his responding to the requirements of a particular role. Of course, personality and role cannot be entirely separated, for an individual's expression of his personality will be influenced by the role he holds and his perception of the requirements of his role will be influenced by his personality. However, we can say that an individual's need dispositions – what he wants from particular situations in which he finds himself – must influence his attitudes to his role. Similarly, his role expectations – how he thinks he should behave in the role and the extent to which he is prepared to subordinate personal needs to group interests – will also influence his behaviour in the role. It can be suggested that the most important integrating factor between personal need dispositions and the role expectations applied to the individual by other groups is *shared values*. The individual is more likely to subordinate some of his personal needs to the needs of the larger group if he believes strongly in the validity of the group's expectations.

Parsons provides us with three ways of looking at values. First he talks about symbol systems which are found in different societies and groups. These he categorises as *systems of ideas* directed at providing ways of understanding what is taking place or should take place in a society or group. *Systems of expressive symbols* associated with helping a society or group to respond to those things which provide gratification and pleasure, and *systems of standards of value orientation* which assist people to evaluate situations, to make choices and to understand when a thing is "right" or appropriate in terms of the values and needs of the society or group.

Parsons' second approach to values is through what he calls *types of standards of value orientation*; the different ways in which people apply their values to the interpretation of situations. He identifies a *cognitive* orientation directed at gaining understanding and answering the question "what is the meaning of this situation for me?" He identifies an *appreciative* orientation related to feelings and answering the question "do I like or dislike this?", and he identifies a *moral* orientation in which the individual makes a moral choice and answers the question "if I do this will it fit in with what I think is right?"

It can be suggested that any successful organizational change requires a cognitive, appreciative and moral response from all those affected in which they *understand* what is happening, *like* what is happening and *agree with what is happening*.[1] Any change in a work situation must therefore fit with the need dispositions of the

---

1. In some situations it is likely that there is a trade-off between these three responses. For example, liking something may compensate for not understanding it.

individual, what he seeks from the work situation, and the *fit* must be evaluated in cognitive, appreciative and moral terms. It must also fit with the role expectations of the individual, how he defines his work duties and responsibilities, and again the *fit* must be evaluated positively in cognitive, appreciative and moral terms. But it is not enough for the change to be understood, liked and agreed with by those on the receiving end. For the change to work the organization's side of the contracts must also be fulfilled and the behaviour of employees must meet the role expectations of the firm. Management will evaluate the success of any technical change in terms of questions such as "do our employees understand why we are making this change and what we expect of them if it is to be successful?" "Are they motivated and enthusiastic about the change?" "Do they share our values and agree with us that this is the right kind of change to make?"

Parsons' third way of looking at values is related to the next question which management is likely to ask. This is: "are our employees behaving in the way we want them to behave – are they effective producers?" Parsons talks about *types of orientation to action* – a tendency for people to behave in one way rather than another when presented with a certain kind of situation. Here he identifies three types of action. Intellectual or instrumental action where the individual applies cognitive values and logical thought processes to discovering the most efficient means for arriving at a given goal. *Expressive* action where appreciative standards of liking or dislike are used to decide whether or not it is appropriate to want or like a given object, and *moral* action where problems are evaluated and solved-using moral standards [Parsons 1951: 75].

In this paper we are interested in a method for identifying value orientations and so we shall concentrate on Parsons' cognitive, appreciative and moral classification. In the research situations we shall of course be testing out value orientations by looking at behaviour as illustrated by the way new computer systems are designed. Parsons' instrumental, expressive and moral action classifications would then prove of value.

Parsons points out that different societies, groups and individuals will place different degrees of emphasis on the cognitive, appreciative and moral aspects of life.[2] Thus some groups will be characterized by a primacy of cognitive interests and profit-making organizations would be an example here. Other groups will be characterized by a primacy of what Parsons calls cathectic interests, the appreciative or affective aspects of life which are associated with expressive symbols such as art forms and styles. Artistic colonies or hippy groups might be examples of this kind of emphasis. Other groups will be primarily interested in the evaluation of alternatives and in considering the implications of choosing one alter-

---

2. Parsons has a tendency to use different words to describe essentially the same things. The *affective* elements of an individual's personality are related to appreciative or cathectic values. Those things which the individual likes and values because they give him pleasure. *Evaluation* is related to moral values. The evaluation that is made before deciding that something is right or wrong in moral terms.

native as opposed to another. These groups will have strong ethical beliefs and value orientations. Some religious and political groups would fit this category. An interesting problem for the researcher is to identify the strength of each of these orientations and to ascertain the systems of ideas or beliefs, of expressive symbols and of value orienation that exist in an organization.

A problem we are particularly concerned with in this research is that of *integration* – the ability of the individual to integrate with the organization on the basis of some set of shared values. It is now useful to return to Parsons' pattern variables. It will be remembered that he describes these as five discrete choices which everyone must make before he acts. At an individual or personality level they may be habits of choice in that the individual has a tendency to choose a certain way in a particular situation. At the group or collectivity level they are aspects of role definition; that is, they cover the definition of the rights and duties of members of a group, these definitions specifying actions and a particular pattern of choice. At the society or cultural level they are aspects of value standards which are translated into rules and recipes for action. The pattern variables therefore can show alternative preferences, predispositions and expectations.

Let us now examine each pattern variable once again from the point of view of first the individual as an employee and second of a group or association similar to a profit-making organization and establish if this helps us towards a framework for studying individual and organizational values. It must be stressed here that this is now the author's interpretation and not that of Parsons.

1. *Affectivity—Affective neutrality*
From the personality or individual point of view
    Affectivity is a need disposition to take advantage of an opportunity for *immediate gratification* in a particular situation

    e.g., to obtain higher pay   (cognitive response)[3]
        to vent agression,
            to evoke friendship   (cathectic response)
        to pass judgement   (evaluative response)

    Affective neutrality is a need to weigh up situations and choose not to seek immediate gratifiction as this may damage the fulfilment of some long-term need

    e.g., to accept lower pay because this
           will help future promotion
           possibilities   (cognitive response)
        to keep anger under control in the
           interests of future relationships (cathectic response)

---

3. A cognitive response is an intellectual response; a cathectic reponse is an emotional response related to a desire for gratification; an evaluative response is a moral response involving the choice of a course of action.

to refrain from moral judgement in
　the interests of long-term moral
　interest　　　　　　　　　　　　(evaluative response)
From the group or organizational point of view
　　Affectivity is the role expectation that an individual (role incumbent) can re-
　　spond to situations as he wishes and secure an immediate personal gratifica-
　　tion if he desires this. He does not need to resist in the interests of group
　　discipline
　　Affective neutrality is the role expectation that an individual will refrain from
　　striving for personal gratification in the interests of group discipline

2. *Self-orientation – Collectivity-orientation*
From the personality or individual point of view
　　Self-orientation is a need to pursue a given interest or goal without regard to
　　group interests
　　Collectivity-orientation is a need to take into account shared group values which
　　the individual feels obliged to attempt to realise
　　e.g.,　taking part in a strike would evoke cognitive, cathectic and evaluative
　　　　　considerations
From the group or organizational point of view
　　Self-orientation is the role expectation that it is permissible for a group member
　　to look after his own interests
　　Collectivity-orientation is the role expectation that a group member is obliged to
　　take account of the values and interests of the group

3. *Particularism—Universalism*
From the personality or individual point of view
　　Particularism is the need for the individual to respond to a situation in terms of
　　his own perception of the requirements of that situation
　　Universalism is a need to respond in conformity with some general standard
From the group or organizational point of view
　　Particularism is the role expectation that when faced with a choice, priority
　　will be given to an individual's own special qualities
　　e.g.,　when selecting a new employee
　　　　　when considering someone for promotion
　　Universalism is the role expectation that when faced with a choice priority will
　　be given to universal standards
　　e.g.,　selection will be restricted to docker's sons
　　　　　promotion is only on the basis of length of service

4. *Ascription – Achievement*
From the personality or individual point of view
　　Ascription is the need to respond to an object (another person) because of the
　　nature of the object (his attributes such as character and personality) rather
　　than its (his) past, present or future performance

31

Achievement is the need to respond to a social object because of its performance not its attributes (it is what he does not what he is that matters)

From the group or organizational point of view

Ascription is the role expectation that the individual when relating with others will give priority to attributes, not performance

Achievement is the role expectation that the individual when relating with others will give priority to actual or expected performance

5. *Specificity – Diffuseness*

From the personality or individual point of view

Specificity is the need to respond to a particular kind of situation in a contained and uniform manner; the same kind of situation always evoking the same kind of response

Diffuseness is the need to respond to the same kind of situation in many different ways depending on other factors operating at that time

From the group or organizational point of view

Specificity is the role expectation that an individual (role incumbent) will respond to the same kind of situation always in the same way. He will have predictable, patterned responses (Go by the rule book)

Diffuseness is the role expectation that an individual will respond to a situation in many different ways depending on the other factors operating at that time

If one accepts that the average industrial organization will require of its employees *affective neutrality,* a *collective orientation, universalism*, an achievement orientation and a *specific* set of responses to the work situation then we can say that it is looking for *discipline,* a shared set of *values, conformity* to agreed standards and procedures, an emphasis on *achievement* and a set of *uniform* patterned responses from employees that fits the requirements of the tasks and technology. We have three rather different kinds of values here, for discipline, conformity and value consensus are required attitudes to the organization as a whole, an emphasis on achievement is a required role orientation which the individual applies to himself and to others, and specificity implies an acceptance of a particular kind of role structure. We are therefore concerned with role attitudes, role orientation and role structure. The problem for the researcher is to establish why the organization requires this kind of behaviour and here we must examine both the environment in which the organization operates and the values of its top management, who are seen as having an important role in shaping institutional values.[4] We also need to know how different groups define acceptable behaviour in each of the value

---

4. Top management values are likely to be a product of both personality and role. Organizational values may be greatly influenced by the beliefs and dispositions of powerful men at Board level. They may also be influenced by the nature of the goods and services which the organization has been set up to provide and the influence this has on top management role responsibilities and expectations.

categories listed above, and those areas in which the organization requires a strict adherence to its value system compared with those where it is willing, or forced, to accept a lack of conformity.

As an example, let us now examine the difficulties of getting a good value *fit* when the traditional profit-making organization recruits the "liberated" individual who has a strong interest in achieving his personal self-development needs within the work situation (Table 1).

Table 1. An example of a problem of *fit* between organizational and individual values

Role expectations of organization (traditional, profit-making)	Need dispositions of individual (liberated, self-developing)
Ethical (shared values) Employees are required to be identified with, and share, organizational values (universalism	*Ethical (personal values)* Individual wants to respond to situation in terms of his own values particularism)
Compliance The organization requires discipline and obedience and for priority to be given to organizational interests (affectivity neutrality	*Non-compliance* The individual wants, when the opportunity arises, to secure some immediate personal gratification through looking after his own interests affectivity)
Conformity The organization wants acceptance of, and conformity with agreed norms, standards and procedures (collectivity-orientation	*Non-conformity* The individual wants to develop himself, express himself in his own way self-orientation)
Performance The organization wants performance to be valued more than personal qualities (achievement	*Person* The individual wants to be valued for his attributes as a person. What he "is" not what he "does" ascription)
Task The organization wants an acceptance of existing technology and task structures. Specific responses to specific work demands (specificity	*Task* The individual wants a set of tasks that fit his needs for variety, autonomy, etc. diffuseness)

Here then is an example of two sets of values that are in opposition to each other and it seems probable that this sort of individual would be unable to work successfully in this type of cultural environment. Nonetheless it is clear that individuals of this kind are to be found in organizations of this kind and it is interesting to examine the conditions under which such an apparently incompatible relationship can work. It can be suggested that if the individual has a powerful instrumental relationship to work – for example, at certain times in his life he has a pressing need for high earnings, then he may be prepared to suppress his personal needs in the interests of this strong financial requirement. He then enters into a predominantly utilitarian or instrumental relationship with the organization. Again, the individual in his private life may be isolated and lonely and have a great need for social relationships which he can only find in work. He then enters into an affective relationship with the organization and suppresses his other needs in order to achieve this. Lastly, he may have a burning mission to live his life in terms of a strong set of social ethics and may perceive joining a particular kind of organization as the best means for doing this. He will then suppress his personal needs in the interests of his ethical goals and will have a moral, evaluative relationship with the organization. This kind of relationship is more likely to be found in a monastery or a welfare organization than in industry.

If we view work relationships as a series of, often hidden, contracts between management and employees, a successful contractual relationship in terms of values is when the organization and the individual implicitly come to the kind of agreement and understanding as shown in Table 2.[5] The responses of the individual are in *cognitive, appreciative* and *moral* terms, represented as C., A. and M.

The important integrating contract here is the ethical contract. It is the organization's ability to secure a consensus on values that conditions its success in achieving a good fit on all the contractual areas. When attempting to establish the goodness of the fit on these five value contracts, it is important for the researcher to ascertain the nature of the individual's cognitive, appreciative and moral responses to each contractual area and to try and ascertain the relative weight of each kind of response. In some areas of work the cognitive response may be the one that is most important to the individual, in others it will be the appreciative or the moral.

So far we have been considering the organization and its relationship with the individual employee. It will be apparent, however, that an important intervening variable between the individual and the organization (represented by the top management value position) is the individual's own work group. This will make its own demands and establish its own implicit contractual relationship with individual members; although these demands are likely to be viewed as legitimate and willingly accepted by those who value group membership. Thus the individu-

---

5. We are here using the term contract in the Weberian sense of "status" contract; that is a voluntary agreement for the creation of a continuing relationship. P. Selznick, *Law Society and Industrial Justice* (New York: Russell Sage Foundation, 1969), p. 54.

Table 2. Successful contractual relationship on values

The organization asks		The individual replies
Ethical contract * Will you share our value system on matters which we regard as important? In return we will not contravene your personal value system	C. A. M.	*Yes,* providing I understand your values and the reasons you hold them I am still able to act in terms of my important personal values Your values do not contravene my values
Compliance contract Will you accept the amount of discipline which we think is necessary to achieve organizational goals and put the most important of the organization's interests before your own when it is essential to do this for organizational survival?	C. A. M.	*Yes,* providing I understand the reasons for the discipline Those things I really want to achieve quickly, I can I agree with how you enforce discipline
Conformity contract Will you accept our ways of doing things – the procedures and methods that we feel best achieve organizational goals?	C. A. M.	*Yes,* providing I understand their logic They do not inhibit my ability to achieve pleasure from work They fit my personal values on how things should be organized
Performance – person contract Will you accept the amount of emphasis which we put on both performance and personal qualities such as trust, integrity, friendship, helpfulness?	C. A. M.	*Yes,* providing I understand your emphasis and the reasons The emphasis is right in terms of my needs as a person The emphasis is right in terms of my moral judgement
Task contract Will you accept the way we use technology and the way we structure individual and group tasks, providing we make every effort to produce a working environment and a set of tasks that fit with your psychological, financial and security needs?	C. A. M.	*Yes,* providing I understand the constraints in your situation which limit your ability to meet my needs My job is structured in a way that meets my important psychological needs I am not required to undertake anything which offends my personal values

* The ethical contract is placed in a box because it is seen as the dominant contract. Unless there is agreement on shared values, the other contracts may be difficult to realise.

al who takes a strong moral position on joining a union or coming out on strike may find that he has to subordinate his own moral position to the values of the group or else create an intolerable personal situation for himself. For the research-er an understanding of values requires a knowledge of how individual values deviate from, or overlap with, the values of the work group. To what extent has an individual to modify his personal values if he wishes to exist comfortably both with his work mates and with management.

The organization (top management) will also be defining its role expectations by thinking through its cognitive, appreciative and moral needs, influenced in this by its external and internal environments and the kinds of pressures it is receiving from these. It may ask itself some of the following questions.

Do we want our employees to subscribe to our values? If so, what kind of company philosophy will help us to achieve this?

Do we want a disciplined workforce, doing what it is told and putting com-pany interests first? Is it possible to achieve this, if so, how?

Do we want tight procedures and clearly defined methods, if so, how do we achieve this?

Must we judge people on their performance rather than on their personal qualities? If so, how can we motivate people to respond in performance terms?

Do we want employees to work on tightly specified tasks with few areas of discretion? If so, can we find employees who will tolerate this situation?

By asking these kinds of value questions the organization may have second thoughts about setting itself up in this tightly controlled way, or, if it believes it has no alternative, then it will recognise that it will have to give a great deal of hard thought to meeting the employees side of the value contract and to making its own values acceptable.

The researcher may find it useful to attempt to place organizations or organiza-tional units such as departments or sections on a continuum such as shown in Table 3.

Information relevant for making a judgement on where an organization or department lies on this continuum can be obtained from interviews, but it can also be obtained from documents. For example, attempts to achieve *shared values* can sometimes be identified from statements on company philosophy, from per-sonnel policies, and communication and consultation strategies. The emphasis on *discipline* can be ascertained by looking at disciplinary policies, supervisory train-ing procedures, etc. The emphasis on *standard methods* and procedures can be as-certained by examining the number of standard operating procedures and the sanctions for deviance. Concern for *productivity* will show up in evaluation pro-cedures such as M.B.O. and the kind of targets that are set and the nature of the monitoring activities. A tight *task structure* can be identified by observing jobs and collecting formal job descriptions.

Another area of interest to the researcher is the strategies used by management to get its values accepted. These too can be categorised as cognitive, appreciative

Table 3. Organizational value continuum

Controlled organization	Permissive organization
The extent to which the organization aims for	
Ethical values – a belief in shared values – an agreement on organizational goals and the means for achieving these	a situation permitting a multiplicity of values to exist and be tolerated*
— — — — — — —	
Compliance values – a belief in strict discipline with employees required to put company interests first	loose discipline, with employees able to meet their own needs in work
— — — — — — — —	
Conformity values – a belief in specified procedures and uniformity of methods	people making their own judgements, working out their own methods
— — — — — — — —	
Performance/person values— a belief in an emphasis on efficiency and high production	an emphasis on personal qualities
— — — — — — —	
Task values – a belief in tightly structured tasks with few areas of discretion	unstructured tasks, with many areas of discretion
— — — — — — —	

* In this kind of organization there is often a belief that the pursuit of diverse individual goals assists the attainment of organizational goals. Creative research laboratories might be an example.

and moral strategies. A cognitive approach is one in which management attempts to explain the logic of an action it is about to take and to demonstrate the usefulness of the proposed change to those on the receiving end. An appreciative approach is to stress the beauty and attractiveness of the new equipment and the pleasure that employees will have in the pleasant new environment that is about to be created. A moral approach is one in which the responsibility of the organization to its employees is emphasized. ''We will consult with you and get your views''; or ''we will look after your interests and see that you do not suffer adversely from the change'' may be examples of the approaches taken. Here differences in moral values are likely to show up. The paternalistic firm may be sincere in its attempts to avoid any hardships for its employees but it may not see consultation as part of the ethical contract.

37

Lastly, it must be stressed again that although this section has set out a suggested framework for examining values, the reasons why certain organizations hold certain kinds of values can only be understood by looking at their purposes, environments and histories.

3.3. A FRAMEWORK FOR THE STUDY OF JOB SATISFACTION

In this study we are attempting to gain some understanding of how values influence technical change and the responses of employees to this kind of change. Because computer systems have a major impact on the structure of work and the ways in which jobs are organized we must enter the field of job satisfaction, for we are hypothesizing that there is a relationship between work structure and job satisfaction. In this section we will therefore look at the concept of job satisfaction and develop an analytical framework for the study of this which emerges logically from the value framework set out above.

Any study of job satisfaction requires a clear definition of the meaning of the term and this is hard to find in the management literature. There are many broad generalizations about the nature of job satisfaction but few that are useful operationally [Mumford 1972: 48]. The definition used in this research was derived from Parsons' concepts of "need dispositions" and "role expectations"; need dispositions being defined here as those things which an employee seeks from his work situation and which provide him with feelings of well-being and satisfaction. Role expectations relate to how an employee defines his own role in a work situation, but more importantly it covers the set of expectations that an organization has concerning the attitudes and behaviour of employees in their various work roles. Parsons' concepts therefore lead us to define job satisfaction in two ways. First in terms of the "fit" between what an organization requires of its employees (its role expectations) and what the employees are seeking from the firm (their need dispositions). A good fit on these should lead to what we shall call "mutually beneficial relationships" and these are, in essence, contractual relationships. Second, in terms of what an employee is seeking from the firm (his need dispositions) and what he is receiving (the extent to which his needs are being met), this last fit being assisted or constrained by the requirements of his work role. A good fit on job needs and role requirements should produce employee "job satisfaction".

It will be recalled that Parsons' pattern variables of affectivity – affective neutrality; self-orientation – collectivity-orientation; universalism – particularism; ascription – achievement, and specificity – diffuseness broadly cover the following choice situations:

1. between seeking immediate gratification *or* deferring gratifiction until a future date;
2. between seeking to further interests private to oneself *or* interests shared with others;
3. between deciding to accept generalized standards in the interests of conformity *or* to seek for acceptance of individual differences and a unique approach which

38

may be a response to emotion rather than intellect;

4. between evaluating people and things because of what they are – their attributes, *or* because of what they do – their achievements;

5. between choosing to react to a person or a situation in a widely differing manner according to the perceived requirements of the situation *or* reacting to a situation in a limited and specific way. For example an individual may seek a variety of satisfactions from work or require satisfaction only on the earnings variable – a fair day's wage for a fair day's work.

But phrased in this way they are imprecise and need clarifying before they can be used operationally to determine whether "mutually beneficial relationships" and "job satisfaction" exist in particular situations.

A first attempt at this simplification led to the following set of definitions [Mumford 1970: 71]. These are related to an organization's role expectations – the behaviour which it expects from its employees in the roles to which they have been allotted. And to the individual's need dispositions – what he himself wants from his work situation.

1. *Organizational role requirements – Personal job requirements*
What the organization wants from the individual and what the individual wants from the organization in terms of attitudes and behaviour. Some of these needs will be urgent and immediate, others will be deferred. That is, the organization and the individual will hope to achieve a number in the short term and others in the long term.

2. *Organizational interest – Self interest*
The extent to which an organization expects its employees to identify with its interests and to forego their own. The extent to which an employee wishes to pursue his own interests in the work situation.

3. *Uniformity – Individuality*
The extent to which an organization's objectives cause it to introduce uniform policies, methods and standards to which its employees must conform. The extent to which an individual wishes to behave in a unique and individual way (to express his own individuality, "do his own thing") and seeks a work situation which allows him to do this.

4. *Performance – Personal quality*
The degree of emphasis the organization places on performance as opposed to social and character qualities. The extent to which an individual wishes to be recognised for what he is, as opposed to what he does.

5. *Work specificity – Work flexibility*
The degree of work specificity which arises from an organization's technology and structure. The degree of work flexibility which an individual requires to match his skills, knowledge and personality.

We are now starting to approach a useable framework with which to examine both the contractual relationships between an organization and its employees, and the job satisfaction "fit" between individual needs and the extent to which these are realised in the work situation. The concepts are made more precise in the categories in Table 4.

Table 4.

	The organization	The employee
The KNOWLEDGE contract	Needs a certain level of skill and knowledge in its employees if it is to function efficiently	Wishes the skills and knowledge he brings with him to be used and developed
The PSYCHOLOGICAL contract	Needs employees who are motivated to look after its interests	Seeks to further interests private to himself
The EFFICIENCY contract	Needs to implement generalised output and quality standards and reward systems	Seeks a personal, equit-able effort-reward bargain, and controls, including supervisory ones, which he perceives as acceptable
The ETHICAL contract	Needs employees who will accept its ethics and values	Seeks to work for an employer whose values do not contravene his own
The TASK STRUCTURE contract	Needs employees who will accept technical and other constraints which produce task specificity or task differentiation	Seeks a set of tasks which meets his requirements for task differentiation, e.g., which incorporate variety, interests, targets, feedback, task identity and autonomy

All of these contracts hold closely to the Parsonian pattern variables except that the extremely vague "affectivity – affective neutrality" has been simplified into the knowledge contract. The assumption here is that the way knowledge is ac-quired, used and developed is an important factor both in an organization's effi-ciency and in the job satisfaction of its employees. It is also an area which contains a great deal of deferred gratification or affective neutrality, for many individuals are prepared to defer financial rewards in order to achieve greater intellectual satisfaction at a later date through the technical and professional qualifications which they spend time acquiring.

If we examine the organization's side of the contractual relationship, it is ap-parent that many factors in an organization's environment will influence how it defines its needs. These will include its product market and the kinds of pressures this exerts; its technology; the labour market in which it operates and the ability of this to supply required job skills; local culture, and various national influences such as legislation [Mumford 1972: 51].

If we turn now to the job satisfaction fit between job needs and work role requirements, the usefulness of Parsons' pattern variables is confirmed by an examination of the literature on job satisfaction. Five different influential schools of thought are to be found there, each of which emphasizes that a particular, and different, set of variables has a major influence on job satisfaction. Each also appears to have a relationship to one of the Parsonian pattern variables. First there is what can be called the *psychological needs* school; those psychologists such as Maslow, Herzberg, Likert, etc., who see a relationship between the fulfilment of psychological needs and work motivation and believe that the two combined produce feelings of job satisfaction. They focus on the needs of individuals for such things as self-development, achievement, recognition, responsibility and status [Herzberg 1966; Likert 1967; Maslow 1954]. A second school devotes its attention to *leadership* as a factor in job satisfaction. Psychologists such as Blake and Moutin and Fiedler see the behaviour of supervision as an important influence on employee attitudes and they therefore direct their attention to leadership style and the response of subordinates to this [Blake and Mouton 1964; Fiedler 1967]. All of these ideas can be linked with Parsons' first two pattern variables: affectivity – affective neutrality and self-orientation – collective orientation and with our *knowledge* and *psychological* contracts.

A third school, represented by Lupton, Gowler and Legge, approaches job satisfaction from quite a different angle and examines the effort-reward bargain as an important variable [Gowler and Legge 1970; Lupton and Gowler 1969]. This leads to a consideration of how the wages and salaries of particular groups are constructed and the influence on earnings and attitudes to these of factors such as over-time pay and the perceived state of the labour market. Some psychologists maintain that people have a subjective perception of what is a fair day's work. They believe that if this is not obtained then job satisfaction will not be high [Jaques 1961]. Wages and salaries form part of an organization's control system; they are the means through which management ensures that it gets the necessary skills and the level of output it requires. If they are included with other control systems such as production, financial and supervisory then we have a category which is not too far removed from Parsons' universalism – particularism pattern variable which we initially redefined as uniformity versus individuality, and later as the *efficiency* contract.

Yet another school of thought approaches job satisfaction from an entirely different angle and sees management ideology and values as having an important influence on job satisfaction [Crozier 1964; Gouldner 1955]. For example, certain organizations may place a great deal of importance on recruiting staff who have certain social qualities – perhaps those inculcated by public schools – while others are more interested in drive and high performers. This approximates to Parsons' ascription – achievement pattern variable and to our early performance versus personal quality category. If widened to include value judgements on other aspects of behaviour and relationships within the work situation it is covered by the *ethical* contract. Today this contract would also include democratic behaviour and the distribution of power [Mumford 1972b: 124].

Lastly there are social scientists who claim that all the aspects of the work situation described above are extrinsic to the tasks an employee has to carry out and therefore a less important factor in job satisfaction than the work itself and the way this is structured. This group concentrates on the *content* of work and on job design factors. (There are many well-known names associated with this school of thought including [Cooper 1974; Davis 1972; Emery 1969; Herbst 1974; Thorsrud 1969].) Because job design is greatly influenced by technology and the amount of discretion in most jobs is influenced by the nature of the technology associated with them, this school of thought fits Parsons' specificity – diffuseness pattern variable which we first redefined as work specificity versus work diffuseness. As we see this as being largely determined by the structure of tasks, we now call it the *task structure* contract.

The relationship between an employee's job needs and the requirements of his work role is better defined as a "fit" than as a contract and so we shall reserve the term contract for the normative "mutually beneficial" relationship. We can now set the job satisfaction fit out as in Table 5.

I am indebted to a fellow social scientist, Dan Gowler, for drawing my attention to the fact that this job satisfaction model can now be subdivided into three broad categories. The *knowledge* and *psychological* fits are related to needs which are associated primarily with the *personality* of the individual. The knowledge fit is cognitive in character and a product of a desire to learn and understand. The *psychological* fit is affective or cathectic and associated with feelings of pleasure or displeasure, liking or dislike [Gowler 1974]. These psychological needs must be met if the individual is to achieve good psychological health.[6]

In contrast the *efficiency* and *task* fits describe the characteristics of the in-

---

6. The reader may now like a reminder. In the area of emotional needs and responses Parsons uses a number of different words which tend to cause confusion. These are *affective, cathectic, appreciative* and *expressive.*

Defined very simply, they have the following meaning:

Affectivity is a *mental state* in which people emotionally want to take advantage of an opportunity for pleasure; they do not rationally weigh up the consequences of doing this.

Affectivity is accompanied by a *cathectic* orientation, or a tendency to take some action which will lead to the attainment of the desired gratification. (Cathexsis is defined as the attachment to objects which are gratifying and the rejection of those which are unpleasant.)

Cathectic problems such as "do I like this painting or not?" are solved using *appreciative* standards which assist *judgement* and *interpretation.*

Systems of ideas or symbols which help people to form opinions on what is beautiful and pleasant are called *expressive.*

When it is decided to act the most pleasant way of doing this will be *expressive* in contrast to *instrumental* which is the most efficient way.

Table 5. Job satisfaction

	The employee's job needs		The requirements of his role
The KNOWLEDGE fit	He: wishes the skills and knowledge he brings with him to be used and developed	A good fit exists when	He: Believes that his skills and knowledge are being used and developed to the extent he wishes
The PSYCHOLOGICAL fit	Seeks to further interests private to himself, e.g., to obtain such benefits as: achievement, recognition, responsibility, advancement		Believes that his private interests are being successfully catered for
The EFFICIENCY fit	Seeks a personal equitable effort-reward bargain and controls, including supervisory ones, which he views as acceptable		Believes that financial rewards are fair and other control systems acceptable
The ETHICAL fit	Seeks to work for an employer whose values do not contravene his own		Believes that the philosophy and values of his employer do not contravene his own values
The TASK STRUCTURE fit	Seeks a set of tasks which meets his needs for task differentiation, e.g., which incorporate variety, interest, targets, feedback, task identity and autonomy		Has a set of tasks and duties which meet his needs for task differentiation

dividual's role, normally determined for him by the organization. A state of job satisfaction implies that there is a good fit between the requirements of the occupational role and the personality needs of the individual. Where a good fit does not exist it can be achieved either by the role occupant manipulating the role so as to make it better fit his personality needs, or by his adjusting his personality needs to the constraints of the role. For example, people with a strong achievement orientation may be able to sublimate this if opportunities for achievement are not to be found in a particular occupational role and either the state of the labour market or personal circumstances prevent a change of job.

The *ethical* fit is the factor which *integrates* these personality and role requirements. The values of the employing organization will influence the extent to which it is sympathetic and sensitive towards the personality needs of the individual and makes an attempt to structure both occupational roles and the total work situation in such a way that these needs can be met. Similarly if there is a good fit between how the employee thinks he should be treated and how he *is* treated by his employer then he is likely to conform more easily to the requirements of his role. The ethical fit also acts as an adjustment mechanism between the requirements of personality and role. A set of values which is common to both the employee and the employing organization may lead to both parties being prepared to tolerate an imperfect fit on the other variables. Employees may be willing to undertake jobs which they do not like very much and which do not give them intellectual stimulation because they strongly support the values on which the purpose of the organization is based. The social services or nursing provide many examples of this kind of adjustment. Similarly, an organization may be prepared to accept employee behaviour of which it does not approve – unofficial ways of getting things done, for example – because it appreciates that both staff and the organization are striving to achieve a common purpose. The revised job satisfaction model can therefore be depicted as shown in Figure 1.

Figure 1. Personality needs – role requirements

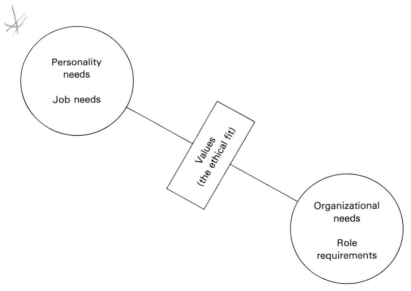

Personality needs are not the only factor influencing job needs, for the latter will also be affected by factors such as an individual's perception of the state of the labour market at any one moment in time. If it is tight he may lower his expectations and accept a job that is less skilled or less well paid than he would ideally desire. Job needs and expectations can also be influenced by opinion derived from such influences as professional literature or professional associations, or from custom and practice, or from school and university, on what an individual can legitimately expect from particular kinds of work situation. All of these, and many other factors, affect the way his innate characteristics are expressed as a set of needs in the work situation. It has to be recognised here that we are defining job needs as something that an individual thinks he wants, rather than what he may really need if he is to develop his full psychological potential.

Let us now examine these personality needs in more detail, once again using Parsonian theory. Parsons suggests that there are three kinds of orientation which people use when confronted with new situations. There is a *cognitive* orientation related to understanding the situation – its requirements and possibilities, a *cathectic*[7] orientation which produces an appreciative or emotional response to the situation, and lastly, a *moral* or evaluative orientation which leads to a judgement that the situation or parts of it is right or wrong, good or bad. Certain personalities or certain situations will evoke responses which are predominantly cognitive; others will produce responses with a strong emotional content, and others will stimulate responses in which moral or evaluative judgements predominate. All of the "fit" areas can be considered in terms of these three orientations. It becomes possible to establish if particular personality needs are based primarily on a desire for meaning and logic; or if they have emotional sources associated with what gives the individual pleasure or discomfort, or if they are a response to moral judgements made by the individual. Similarly, it can be ascertained if the individual is responding to aspects of his occupational role in terms of cognition, emotion or moral judgement. Most people are likely to respond to situations in terms of all three orientations, but certain types of personality and certain kinds of situation evoke a stronger cognitive, cathectic or moral orientation than others. Certain kinds of occupational role may require responses which are cognitive rather than cathectic or moral, or vice versa. For example, the mathematician has a cognitive role, the actor a cathectic role, the priest a moral role.

This idea of cognitive, cathectic and moral orientations can now be applied to the personality variables and the logic of these improved. One thing immediately becomes clear, our knowledge fit relates to cognitive needs for learning, understanding and intellectual stimulation. There therefore appears to be a rational reason for keeping this separate from the psychological fit. The psychological fit as originally conceived was influenced by the work of Maslow and Herzberg and

---

7. The terms affective, cathectic and appreciative all refer to emotional responses related to linking or disliking, feeling pleasure or displeasure.

based on their perception of psychological needs in work, particularly those associated with motivation.[8] The origins of the Herzbergian model appear to be empirical rather than theoretical and his association of the fulfilment of psychological needs with the kind of motivation of which management approves seems particularly dubious. An examination of the ideas of Parsons suggested that there was a better basis for a classification of need dispositions. Psychological needs were clearly related to the attainment of those things which either gave the individual feelings of pleasure and satisfaction or reduced the possibility of his feelings of discomfort or displeasure; they were therefore cathectic in character.

It is now suggested that the psychological fit can be split into two categories and four sub-categories of cathectic needs as follows:[9]

1. *Esteem needs*
1.1 Self-esteem needs related to the individual's *self-image*.
These will cover some of the Herzberg motivators such as a need for achievement and for feelings of responsibility.
1.2 Group esteem needs related to the individual's association with *others* in the work situation.
These will cover a desire for prestige, status, advancement, respect and other indications that the individual is held in esteem by others.
2. *Security needs*
2.1 Security needs related to the individual's *personal security*.
These will cover feelings of physical and psychological safety. An individual's belief in his own competence, in the security of his job and in his ability to earn enough to meet his needs are three examples.
2.2 Security needs related to the individual's security as a member of a *group*.
Covering friendship, pleasant interpersonal relationships, an absence of conflict in the work situation.

The personality variables can be depicted as in Figure 2.

The fit areas can be described in the following manner:

The *knowledge* fit (cognitive needs)	A cognitive requirement related to the employee's need to understand the logic of his job and work situation and to have an opportunity to use and develop his intellectual skills.
The *psychological* fit (affective needs)	An affective need to enhance one's self-esteem through feeling achievement, responsibility, etc., and to secure the esteem of others shown through the awarding of status, respect, promotion, etc.

---

8. It has, of course, been discussed by many other writers, in particular Schein [1965: 11].

9. These, together with the knowledge fit, are close to the need hierarchy model of Maslow. This theory of motivation encompasses *safety needs, social needs, esteem needs* – covering self-esteem and the esteem of others – and *self-actualization* needs.

46

Figure 2. Personality variables

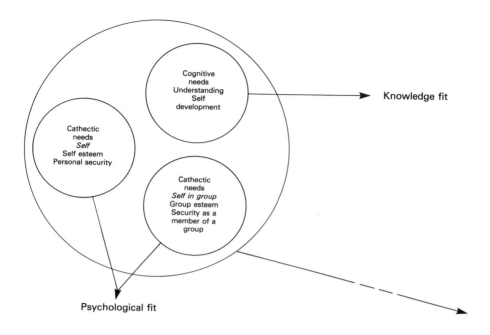

The *psychological* fit (affective needs)   An affective need to avoid feelings of personal insecurity through the provision of a secure job, adequate earnings, physical safety, work competence, etc.; and to achieve security as a member of a group through opportunities for pleasant interpersonal relationships.

Moral or evaluative orientations will form part of the *ethical* fit.

Let us now consider role requirements in terms of cognitive, cathectic and moral orientations in order to establish if the factors considered under the efficiency and task fits can be made more specific. But before doing this there is a need to examine the nature of the efficiency fit. Consideration of this fit area makes it apparent that what is being examined here are the control and support systems which are associated with any set of tasks. We have included work controls, supervisory controls and pay levels and structures which we see as a form of social control, in this fit. Therefore it would seem logical to speak of it as the control fit. But the notion of efficiency also implies that the employee is given the resources necessary for him to do his work effectively, in other words, he requires a support system. It is therefore proposed that the efficiency fit be sub-titled the control/support fit.

Role requirements can therefore be set out as in Figure 3.

Figure 3. Role requirements

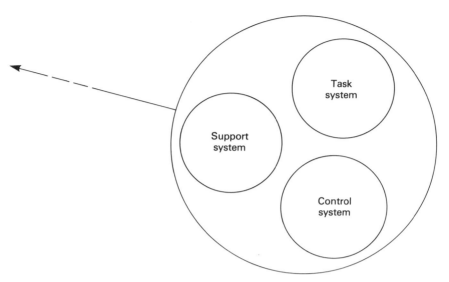

Role requirements

The efficiency fit incorporating both control and support systems, now represents the task environment and can be described as follows:

The *efficiency* fit (control and support systems – the task environment)	The extent to which the *control* systems associated with the work situation, including wage structures, supervision and work controls, assist the employee to perform his job efficiently and agreeably and provide him with incentives rather than constraints.
	The extent to which the employee receives the necessary support services to enable him to carry out his job efficiently, comfortably and without stress. These support services will include equipment, working conditions, training, advice and stability as a feature of the work situation including balanced work loading and work continuity.
The *task* fit (the task itself)	The extent to which the set of tasks for which the employee is responsible are structured in such a way as to provide him with opportunities for varied work of sufficient complexity to enable him to take responsible decisions and broad enough to give him a meaningful job. Technology may act as a constraint on task structure.

48

Gowler has identified the *ethical* fit as the principal integrating factor in our job satisfaction model [1974: 14]. It is a very broad one, covering every aspect of a firm's activity, in particular organizational values on such things as participation, consultation, communication and ownership and control.

Returning to the cognitive, cathectic and moral orientations and their relationship with role expectations, it is clear that ideally the individual should make a positive cognitive, cathectic and moral response to each aspect of his job and work situation. In other words he must understand it and know what is required of him and why this is required. He must also like it, or at least not actively dislike it, and he must agree with it. It should not offend any of his moral principles. He should feel that what is expected of him is right and proper and based on sound judgements. This is the ideal situation which will lead to a state of "perfect" job satisfaction if the *ethical* fit is also good and there is considerable overlap between the values of the individual and those of the organization. Almost certainly it is hardly ever found and many trade-offs must take place with people accepting one kind of gratification in place of another and being prepared to defer short-term gratification in the interests of long-term advantage.

Gowler has pointed out some of the difficulties in achieving this degree of harmony. He says:

> Job satisfaction depends in part upon the individual's ability to achieve a degree of harmony between the affective, cognitive and evaluative elements of his personality. Furthermore this ability influences and is influenced by the degree of *fit* on each of the contracts. This may happen if an employer requires an employee to behave in a manner which does not square with his values or to perform tasks which do not meet with his emotional needs and so on. [1974: 7]

He points out that we can experience different kinds of dissonance. For example, *cognitive – affective* dissonance can occur when our intellectual view of work is challenged by an appeal for an emotional response which does not appear to be based on logic. We are then likely to feel some psychological discomfort such as anxiety or even fear. *Evaluative – affective* dissonance is likely to occur when an individual's moral sense is affronted. He will then feel indignation and anger.

It can also be argued that a state of perfect satisfaction is not only unattainable but also undesirable. It is unattainable because people's needs are not static but constantly change as the individual moves through different stages of his life cycle or recognizes that his situation contains constraints that he must accept and live with as they cannot be altered. At the same time in a period of rapid technological change people's jobs are altering and they are being required to learn new skills and take on new opportunities. It is undesirable because a perfect fit on all the variables, including complete understanding, liking and agreement with every aspect of the work situation would produce complete inertia. People would become so comfortable that they resisted change and impeded necessary progress.

As Gowler has aptly put it what is required is a certain amount of "grit in the fit", so that the individual in working to improve the fit between his own personality needs and the requirements of his occupational role stimulates change that is useful to the organization and which at the same time assists his own self-development [Gowler 1972: 25].

The revised job satisfaction model now takes the form of Figure 4.

It can be described as follows:

The KNOWLEDGE fit	Cognitive needs related to learning and intellectual self-development.
The PSYCHOLOGICAL fit	Affective needs related to esteem and security, both personal and as a member of a group.
The EFFICIENCY fit	The task environment.
The TASK fit	The task system.
The ETHICAL fit	Values. The extent to which there is an identity of values between the individual and the organization. From the employee's point of view these values will include how he wants to be treated as a person and how he wants to be able to treat his associates. They may also involve moral judgements on the organization's internal and external relations.

I have specified some of the important factors which need to be examined under the different "fit" headings. There are others but it is hoped that this analytical framework will provide guidelines on where to place them in the job satisfaction model. Some variables can be placed in more than one fit area depending on how they are being considered. For example, training will come in the efficiency fit when it is viewed as a back-up facility to enable the employees to carry out their job efficiently. It will come in the ethical fit when it forms part of a company philosophy on employee development. Similarly, responsibility can be a psychological need or a task characteristic. If it is the latter then there has to be some way of isolating and measuring this task characteristic. Earnings or pay is the only factor which in the model appears in both the need disposition and the role expectation categories. From the employee's viewpoint the opportunity to earn enough to meet his and his family's needs is an important element in his personal security. From the organizational point of view it is an important means for regulating employee behaviour.

A major deficiency of the model as described so far is that it is solely concerned with the individual employee's relationship with the employing organization. It takes no account of the fact that many in-work needs are met and conditioned by the work group rather than the wider organization. The model therefore requires another element if it is to represent with any accuracy the fit between job expectations and job requirements. Figure 5 provides this.

Job needs are constantly modified through the interaction of the individual with his work associates. For many people the work group will have a need-ful-

Figure 4. Revised job satisfaction model

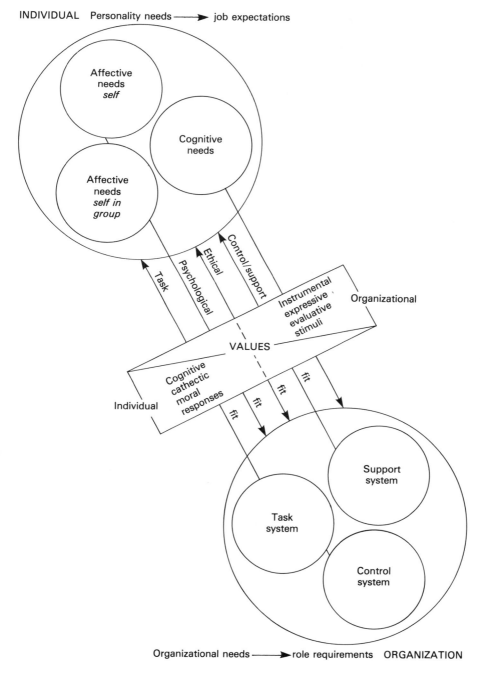

52

Figure 5.

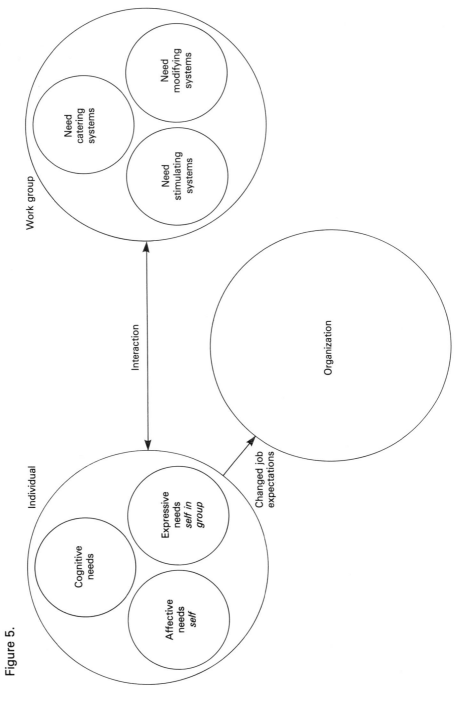

filling function. For example, it is the source of the esteem and pleasant social relationships that most people value in work. Needs are also stimulated by the work group whose members may suggest to the individual that he should expect certain things from the larger organization – higher pay or more interesting work or greater job security, for example. At the same time needs are modified and normative ways of behaving established through the social control mechanisms of the group. All of these things will alter the job expectations of the individual and affect his relationships with the employing organization. Any understanding of job satisfaction requires an understanding of the relationship between the employee and the work group as well as between the employee and the organization.

It is also important to mention another limitation of the model. This is that although it can provide the researcher with a good understanding of the level of job satisfaction in a particular organization and an appreciation of which fit areas are good and bad, it tells us only what people are seeking from work, not why they are seeking these things. Any depth study of job satisfaction requires a journey into individual psychology in order to ascertain why particular individuals have particular sets of needs and values. Similarly, from the perspective of the organization, we can find out what values the organization holds but we need to know why it holds these values and from where they have been derived. Organizational values will determine why the efficiency and task systems are structured in a particular way; they will also determine the extent to which the organization wishes, or is able, to meet the job needs of its employees.

3.4. A FRAMEWORK FOR DESCRIBING JOBS

In order to produce an organization of work that meets human needs and leads to an increase in both job satisfaction and efficiency, it is important to have some knowledge of different ways in which jobs can be structured and to have an analytical framework against which existing jobs can be measured. This section describes the analytical framework used in this study.[10] This framework is derived from current ideas on what employees are seeking from their work situations in terms of job satisfaction and on how task activities can be combined into interesting, challenging jobs [Davis 1972; Cooper 1973; Taylor 1975; Mumford 1975].

An important principle of job design is giving the employee a degree of control over his work environment. In order to ascertain the extent to which an employee has control we therefore need a control model of some kind.

The method of classifying work in control terms set out below has been derived from a cybernetic model developed by Stafford Beer. This defines work as having a number of different control levels [Beer 1970].

---

10. It was necessary to create a new method for describing jobs as existing methods had been developed for job evaluation or training purposes and did not seem relevant to job satisfaction.

Level 1   *we can call the operating* level and this is the set of tasks which the employee has to undertake in order to make something, or process something or record something. Most of our daily activities are concerned with these operating tasks. Wilfred Brown has called them the *prescribed* part of any job. The tasks we must do, as opposed to the *discretionary* parts, the tasks where there is a choice on whether we do them or not [Brown 1960: 102].

Level 2   is the *negative control* or *antioscillation* level that prevents problems or variances occuring and corrects them if they do. It is concerned with the availability of necessary resources to carry out the work and to correct problems if they occur. In traditional work situations, allocating resources and handling problems is often a supervisory responsibility.

Level 3   is the *positive control* or *optimising* level and this is concerned with the coordination of work and with ensuring that the tasks which an employee performs fit smoothly and logically together and integrate well with those of other employees so that the whole department is running smoothly and efficiently. It is therefore primarily concerned with the boundaries between one task and another and one job and another. Again, in traditional work situations, work coordination between employees is a supervisory responsibility.

Level 4   is the *development* level. This concerns those aspects of work, often discretionary rather than prescribed, where a job can be developed by bringing new ideas to bear on it. It covers opportunities to try out new methods or plan new tasks or objectives. This aspect of work requires creativity and initiative. Most management jobs have a level 4 component but many lower level jobs contain little, if any, of level 4.

Level 5   is the *overall control* level. This level is concerned with the total management of the job and the way in which all the levels fit together. In traditional work situations this may be a departmental management responsibility.

In the job structure comparisons shown in chapters 6 to 9 these are the five control levels shown at the top of the sheet.

Most non-supervisory jobs have a large level 1 component with a much smaller element of the other levels or none at all. We therefore need a sub-classification so that it is possible to assess the number of desirable job characteristics contained in the level 1 operating aspects of work. This is derived from the work of Cooper at Lancaster University [1973, 1974]. A job characteristic that is associated with job satisfaction is "variety". This covers both physical variety in the sense of the opportunity to undertake a number of different tasks, and intellectual variety meaning the opportunity to use different levels of skill. It is believed that skill variety is a more influential factor in producing job satisfaction than physical variety. The second and third columns in our job structure comparisons table show the number of tasks and numbers of skills associated with the jobs which have been observed.

Another factor related to job satisfaction is discretion or the opportunity to exercise choice. Here two kinds of choice are available. First what we can call a choice of "means" covering the selection by the employee of the best methods or tools for performing a particular set of tasks. Second there is choice associated with the use of "knowledge" to solve a particular problem and this will cover using judgement and initiative and taking decisions. Means choice and knowledge choice are columns four to seven in the job structure comparisons table.

A third characteristic associated with job satisfaction is job identity – the fact that a job is well-defined, well-integrated, has a progressive aspect so that something important is seen to have been achieved between the start and finish of the job and has a long enough task cycle to enable this significant contribution to the development of the product or service to take place. These aspects of work are evaluated in the next five columns of the job structure comparison table. Jobs are assessed in terms of whether there is an uninterrupted task sequence between the beginning and the end of the job. Many jobs have a hiatus in the middle while information is collected from other departments or while a computer processes input information. Where a computer causes an interruption this is shown by the

symbol ⬚

By task cycle is meant the time that it takes an employee to handle his main work activity. If he is an order clerk then this will be the average time it takes for him to process each order; a task cycle of much less than twenty minutes suggests that a job is very repetitive and routine.

Job relationships are another factor associated with job satisfaction. Most employees like to work as members of groups and this is noted in column 13. Some jobs require considerable interpersonal relationships with other members of the group or with outside groups and this is shown in the next column.

Lastly there is goal clarity and the ease or difficulty of goal achievement. By these are meant, firstly, does the employee understand clearly what he is trying to achieve in his job and the quality or output standards which he should be meeting? Secondly, does the achievement of the goals and objectives associated with his job stretch him enough to provide a challenge while not being so difficult that it causes frustration. These aspects of a job are noted in columns 15 and 16.

If jobs are redesigned by incorporating into them more of the factors which we have been discussing as level 1 or operating activities, this can be called horizontal improvement. If they are enriched by incorporating activities which move them up to higher levels of our control model then this is vertical improvement.

Let us now use this framework for considering some popular approaches to work design.

Many of these are concerned with improving *level 1*, the operating or day-to-day task level, for the employee. This is true of *job enlargement* and *job rotation*.

Table 6. Job structure comparisons

	LEVEL 1 — OPERATIONAL LEVEL							HIGH CONTROL LEVELS				
	Variety		Means choice	Knowledge choice	Job identity	Job relationships	Goal clarity	LEVEL 2 Anti-oscillation		Optimisa- tion LEVEL 3	Develop- ment LEVEL 4	Overall control LEVEL 5
CONTROL LEVELS	Reduce boredom	Give sense of personal control	Give feeling of achieve- ment	Use of initiative	Give sense of making important contribution	Give sense of team work	Give sense of confidence	Problem prevention	Efficiency improvement	Creativity	Autonomy	Key task

Column descriptors (left to right):

- No. of tasks
- No. of skills
- of methods
- of work sequence
- Use of judgement
- Use of initiative
- Clearly defined start to job
- Clearly defined end to job
- Uninterrupted task sequence
- Long task cycle 20 + minutes
- Visible contribution to product or service
- Works as member of group
- Considerable inter-action required
- Clear work objectives
- Objectives not too easy or too difficult
- Can requisition re-quired resources
- Can correct errors, solve problems
- Coordinates own work activities
- Coordinates group work activities
- Can improve methods
- Can improve prod-uct or service
- Individual free from supervisory control
- Group free from supervisory control

Definition of skills:
These are the day to day operating tasks
- A. Communicating in writing
- B. Communicating verbally
- C. Arithmetical skills
- D. Machine operating skills
- E. Checking/monitoring/ correcting
- F. Problem solving
- G. Coordinating
- H. Supervising

Here variety is introduced into the job by giving the worker more tasks to carry out or by allowing him to move round a number of tasks, spending a period of time on each. None of these tasks may require much skill for their performance. This kind of approach can be useful in that it reduces work monotony and will be appropriate if a particular employee group is generally satisfied with work and merely wants more variety in order to have a higher level of job satisfaction. It is unlikely to be adquate for groups whose job satisfaction needs are more complex and related to a desire for opportunities to use skill, meet challenge, and exercise control.

A management favourite at the present time is *job enrichment*. This also tends to focus on *level 1* although the approach is more sophisticated. Work is now designed so that the employee is able to use a number of different skills, some of which are quite complex and require judgement to be exercised, choices made and decisions taken. Many offices where clerks deal directly with customers are able to offer this kind of work. This kind of job enrichment seems to improve both job satisfaction and efficiency in many situations, providing that employees can be trained to the necessary level of competence. It may be difficult to introduce into departments where there are few experienced employees and where there is a very high level of labour turnover. Also because this approach does not give clerks any responsibility for level 4 development activities it does not lead to any new thinking or better ways of carrying out the work.

It is not easy to incorporate level 4 or the development aspect of work into clerical jobs although intelligent staff are likely to find that an opportunity to develop new ideas and try out new methods is one of the most satisfying aspects of their work. If jobs have this component it is also of great advantage to management as work methods will be constantly reviewed and improved and suggestions will be coming from staff for innovations that assist the prosperity of the business. At present many managements try to stimulate these new ideas through suggestion schemes but employees are often reluctant to participate in these and a suggestion coming from one individual may not be acceptable to staff as a whole. Management by objectives has proved a useful method for making *managers* aware that they have level 4 responsibilities and should be directing their attention to new and improved ways for organizing the work of their department, to planning effectively for the future and to increasing the market penetration and profitability of their companies. Here it has been demonstrated that the manager cannot make an improvement at level 4 unless level 2 is also well-organized, that is, he has the resources necessary to allow him to improve methods, and plan and increase business. It may be that something similar to management by objectives could be applied to *employees who are not managers* but a great many new initiatives coming from individual clerks might cause more problems than improvements.

A way of incorporating level 4 development activities into work becomes possible if we stop thinking about individual jobs and turn our attention to autonomous groups or, to use a simpler term, self-managing groups. The focus of our attention is now the group rather than the individual as we are able to consider how we can incorporate the higher levels 3, 4 and 5 into the work responsibilities of a group of non-supervisory staff which work together as a team. If we concentrate our attention on the kind of self-managing group that is multi-skilled in the sense that each member of the group is competent to carry out all the operational activities for which the group is responsible then many things become possible. The group can now take charge of the level 3 coordinating activities and organize its own work so that individual task responsibilities integrate well together and the group works together as an efficient team. At level 4, the development level, a group is more easily able to initiate and try out new ideas and methods than an individual employee. Similarly, if management has great confidence in the ability of its self-managing group then it can hand over a great many of the level 5 or overall control activities to the group. It can, for example, let the group organize its own work activities and set its own performance targets and monitor these (levels 3 and 1). It can give it responsibility for identifying and correcting its own mistakes (level 2). It can give it a budget and allow it to buy its own materials (level 2) and even organize the selling of its own products to customers (level 4). The planning and coordination of all these activities is a level 5 responsibility and the self-managing group is then operating in a manner similar to that of a small independent business within the firm. This kind of group needs no supervisory intervention in its activities and management's responsibility will become one of long-term planning and boundary management. By boundary management is meant ensuring that the work of all the self-managing groups in a department is coordinated and that the work of the department as a whole integrates well with that of other contingent departments.

The self-managing group with responsibility for work levels 1, 2, 3, 4 and 5 is certainly the most revolutionary way of reorganizing work but careful attention must be paid to employee wishes here. The self-managing group may work well in one situation. It may be a disaster in another, and it is very likely to prove unsuccessful if this is not the kind of change that the clerical staff are seeking. They may prefer something less ambitious. The self-managing group that accepts responsibility for work levels 1, 2, 3, 4 and 5 we already find among professional groups, for example in the group practices of lawyers and doctors. It remains to be seen whether it will prove effective with clerks in offices and whether such a form of work organization would be acceptable to many managements.

Nevertheless, Scandinavian experience suggests that the self-managing group is excellent in the right situation and that it provides a stimulating work environment in which staff can really develop their talents [Lindholm 1975]. However, for it to succeed certain things are necessary. First of all, if level 1, the day-to-day activities, is to be improved then the work of the department into which the self-managing groups are introduced must provide scope for multi-skilled work that provides challenge and responsibility. In many offices work has been so strictly and functionally allocated between departments that no re-arrangement of work or creation of self-managing groups can make it much more interesting. In this kind of situation the challenging problem-solving aspect of the work has often been separated off and handed over to a specialist group in a separate department. Any real improvement in work interest can then only be achieved if several departments are merged together thus providing work variety. Secondly, the creation of multi-skilled self-managing groups with the competence to control many of their activities requires intelligent, responsible employees together with excellent long-term training. It may take several years to make all members of a group multi-skilled and if the group suffers from a high labour turnover then management will find the training process an expensive one. Thirdly, the creation of self-managing groups has implications for salary levels and grading schemes. Grading can no longer be related to length of service, it must be related to knowledge and skill. Therefore a new employee who is adept at learning all the jobs for which a group is responsible will reach the highest grade quickly and this may be resented by long-serving members of the firm who have worked their way slowly up a hierarchy of grades over many years. But this is a problem associated with a change from one philosophy of work to another and it should not cause permanent difficulty.

Again increased competence means, in most work situations, an increased salary and the self-managing group may be expensive for a firm because of this. Fourthly, the self-managing group with responsibility for levels 1, 2, 3, 4 and 5 does not need a supervisor and firms may find that they have a problem with redundant first or second line management. Despite these problems many organizations are moving towards self-managing groups and it may be that in a few years time there will be increased pressure from clerks and unions for these kinds of groups. Management too should gain from the increase in job satisfaction and efficiency that seems to be frequently associated with these groups.

3.5. CONCLUSIONS

Before systematic research can be undertaken the investigator requires a set of analytical tools which assist his observations. The components of the analytical framework used in this book have been derived mainly from the work of Parsons. There are four of these, covering organizational values, contractual relationships between an organization and its employees, the job satisfaction fit between employee job needs and the requirements of their work roles and a method for describing jobs based on a cybernetic model developed by Stafford Beer. The theoretical tools described in this chapter may appear rather complex for the simple analysis used in the case studies. These tools are also being used by the author to assist other organizations to diagnose their own needs and those of their employees and to develop new forms of work organization in association with new computer systems.

4. Systems Designers: Their Values and Philosophy

An important objective of this research has been to obtain a picture of the values and philosophy of systems designers and to attempt to gain some understanding of how these values influence the kinds of computer systems which these specialists design. Our hypothesis here is that the design of computer systems is influenced by a vision of man and man's needs and abilities which is derived from the systems designers' own values, training and experience, and that this vision or "model of man" can deviate in a number of important ways from the model of himself and his needs held by a clerk in an office or a worker on the shop floor. Levinson, a psychologist, refers to this as the "great jackass fallacy". He argues that there is an unconscious managerial assumption about people and how they should be motivated. It results in the powerful treating the powerless as objects [Levinson 1974: 70]. It seems probable that the attitudes of systems designers are influenced less by considerations of power than by a strong allegiance to technology, and that technology, and its potential, is a major influence on how they see the world and their own design responsibilities. Systems designers – like all other men – view the behaviour of themselves and others in terms of certain assumptions about human beings, about society, and about men and women in interaction with one another[1] [Phillips 1973: 78].

Simple models of man, if used, would have a number of useful functions for the technically-oriented systems designer. They would enable him to concentrate on the complexity of technology without having to spend a great deal of time considering the complexity of the human being. They would help him to simplify his design process by reducing the number of variables which he would otherwise have to handle. When his systems are finally implemented a discounting of the human factor means that these can be evaluated in terms of their technical rather

---

1. This is also true of the author of this book and as we are about to examine some empirical data on values, and draw some conclusions from this data, it is worth saying a few words about her value position. Like Eric Fromm she believes that "man, not technique, must become the ultimate source of values" [Fromm 1968]. Therefore, although she will attempt to examine the research data objectively, her conclusions are unlikely to be completely "value-free". She would, however, argue that this problem is common to many subjects in the "softer sciences", including medicine and biology, as well as to the social sciences.

than their socio-technical success. But in carrying out such a variety-reducing process the systems designer would, in effect, be creating his own organizational world and the rules by which it will operate. This could turn out to be a phantasy world which bears little similarity to the world of the users of his systems.

Many observers have commented on the values and "models of man" of technical systems designers, although their evidence has usually been derived from examining the human consequences of design decisions and inferring values from these consequences. Little research has been carried out to establish the true nature of the values of the technical specialist. In this chapter we will attempt to establish whether the hypothesis "that the model of man used by systems designers deviates markedly from that held by the users of the computer system" is true for our research situations; including here the clerks, their departmental managers and senior managers with responsibility for these user functions. We will do this, initially, at a general level, aggregating the data from six organizations that were investigated as a preliminary to looking in detail at the values and value "fits" in the four case study situations that are described in chapters 6, 7, 8 and 9.[2] Some comments on how well-known writers and researchers see the values of systems designers are set out below. These kinds of views are today tending to mould the popular stereotype of the technical specialist.

Boguslaw [1965: 202] has said that the "new utopians" (including here computer systems designers) retain an aloofness from the human and social problems presented by the fact or threat of machine systems and automation. "They are concerned with neither souls nor stomachs". He contrasts the approach of the systems designers of the past with that of the systems designers of today.

> The most distinctive characteristic of classical utopian designers is the basic "humanitarian" bent of their value structures. In Sir Thomas More's Utopia, the inhabitants are more concerned with the welfare of their fellow men than with the furthering of their individual fortunes . . . the new utopians are concerned with non-people and with people substitutes. Their planning is done with computer hardware, system procedures, functional analysis, and heuristics . . . . The theoretical and practical solutions they seek call increasingly for decreases in the number and in the scope of responsibility of human beings within the operating structures of their new machine systems.

Louis Davis [1971: 76], one of the major researchers in the field of job design, describes the model of man of the technical systems designer as follows.

> No clear objectives concerning roles for men as men are visible, although objectives are clearly defined for men as machines. When man is considered

---

2. Only four case studies are presented in detail for the following reasons:
1. The author wanted to describe organizations that handled change well rather than badly.
2. Two of the government departments wished their data to remain confidential.

only as a link in a system, design rules do not exist for allocating appropriate tasks to man. Nor are there design rules for designing task configurations to make complete and meaningful jobs . . . . If we turn to how technology is translated into requirements for job designs, we see widespread acceptance for the notion of the technological "imperative" put forth by most engineers and managers. That a substantial part of the technical design of production systems involves social system design is little understood or appreciated.

Sackman [1971: 49] suggests that what he sees as an unwillingness or inability to take account of human factors in systems design stems from the manner in which computers were first used and continue to be used. He says:

The early computers were virtually one of a kind, very expensive to build and operate, and computer time was far more expensive than human time. Under these constraints, it was essential that computer efficiency come first, with people last . . . . The developing computer ethos assumed an increasing misanthropic visage. Technical matters turned computer professionals on; human matters turned them off. Users were troublesome petitioners somewhere at the end of the line who had to be satisfied with what they got.

Sackman continues:

During the formative era in the rise of computers, the message was propagated, with drum-beat monotony, that computer needs are more important than user needs. The user approached the computer room with great trepidation, as if it were a shrine: and he considered himself lucky if the digital deity, as it dispensed its random favours, would service the user's supplication in due order.

This subserviant attitude toward the almighty machine took hold in pioneer designers and users of computer systems and was transmitted to their disciples. Their main energies were directed towards getting working programs and reliable computers rather than understanding and helping people to put information processing to work. The dehumanization of computer services struck deep roots and spread throughout the computer world.

These observations come from writers and researchers who are not unbiased. They all hold strong humanistic value positions and may have an interest in painting a black picture of technologists so that their own more human-oriented approach will appear a valid substitute for the technical approach, and a better fit with today and tomorrow's system design requirements. In effect they are trying to replace technically dominated models or paradigms with socio-technical models.

An alternative vision of the systems designer is put forward by Swanson who sees him in heroic guise.

The designer-hero faces both technical and social obstacles in winning approval for his concept. Mountains of data may require scaling. Micro-seconds of precious computer time within a program subroutine may separate success from failure, and clever algorithms may be required to find one's way within a maze of computational choices. Impossible schedules may be imposed which must nevertheless be met. Organized opposition may be lurking in the wings to "shoot down" one's design plan. And so on.

The initiation of the hero reaches its climax through some decisive victory, a penetration to some ultimate source of power and the securing of its boons. The designer obtains the conclusive conceptual insight, wins "final approval" of his design, and is rewarded with the allocation of those technical and organizational resources needed for its ultimate realization. [1974: 93]

Unfortunately, according to Swanson, the systems designer is unwilling to share his heroic role and so encounters trouble with managers who have to implement his systems, and employees who have to use them. Design is heroic, implementation is dull. The implementor accompanies and facilitates the hero's return, he does not share in the hero's adventures. Similarly, the user has no part to play in the adventure. He may reap some rewards but his role is defined as that of the uncreative, the inactive, the "needy" [Argyris 1971]. Herein lies the crux of the problem, for, as Churchman has pointed out, "the hero is in everyone of us" [1971: 204]. Users too want to share in the adventure and the heroic role.

A major contribution to our understanding of the mind of the scientist, to the nature of scientific developments and to the conditions under which scientific thinking is changed has been made by Kuhn in his book *The Structure of Scientific Revolutions*. Kuhn is discussing professional scientists and the conditions required before major scientific breakthroughs are accepted by the scientific community and allowed to replace old ways of looking at the world, nevertheless his arguments seem likely to fit the mental processes of our systems designers and of those who seek to change them. In the first edition of his book Kuhn argues that "particular coherent traditions of scientific research" take their shape from *paradigms* which he defines as "universally recognised scientific achievements that for a long time provide model problems and solutions to a community of practitioners". These paradigms include "law, theory, applications, and instrumentation together" [Kuhn 1962: 10]. "They are the source of the methods, problem field, and standards of solution accepted by any mature scientific community at any given time. As a result, the reception of a new paradigm often necessitates a redefinition of the corresponding science" [ibid: 102]. In this research we are not concerned with scientific research but with the use of technology. However, we can hypothesize that a process similar to that described by Kuhn is likely to influence the design activities of a system design group. The writers Boguslaw, Davis and Sackman, quoted above, are suggesting that technical systems designers have paradigms which include technology and the design of this, but exclude people and the design of social systems. They, from their value position, are trying to replace these technical paradigms with socio-technical paradigms in which problems are defined in terms of technical *and* human needs and solutions. But Kuhn argues that a new paradigm will only be accepted by a scientific group if the members become aware that their existing paradigm contains some anomalies and does not fit the problem world to which it is applied [ibid: 65]. Our commentators' contribution to a changed approach is to try to make technologists themselves, and those who use or are affected by technology, aware that there is a poor fit between technological solutions and socio-technical problem requirements.

The author too has played this proselytizing role on many occasions, and with some success, as the technical paradigm seems slowly to be giving way to the socio-technical paradigm. Her purpose in this book, however, is not to proselytize, but to analyze and contribute to an understanding of the models of man held by systems designers and users; the extent to which these match or diverge from each other, and how they influence both the design and reception of computer systems.

It must be stressed that the concern of this book is only with the values of a small, specific group of systems designers. The research data should enable us to gain an insight into these values and to trace the design processes associated with the computer systems in our two firms, one bank and a civil service department. In this way we may come to understand how these computer systems have been designed and the reasons for particular design strategies; in particular the influence of certain kinds of values and value judgements on these strategies. No conclusions about systems designers in general will be drawn from the data, although it is hoped to suggest hypotheses which can later be tested on a larger sample.

4.1. ROKEACH VALUES

A good entry point to a study of values is at the meta level of gaining an insight into broader "life" values. A great deal of work has already been done on this subject by Milton Rokeach, Professor of Sociology and Psychology at Washington State University. He has developed two lists of values, one related to terminal or goal-directed values, the other related to instrumental values, or ways of behaving that would assist the achievement of goals. These lists have taken Rokeach a number of years to produce. He began by making a search of the American literature concerned with values, and by talking to colleagues, students and members of the public. The several hundred different values that emerged from this approach were then reduced by eliminating those judged to be more or less synonymous with one another, for example, "freedom" and "liberty", those values which overlapped, and those which were too specific or did not relate to goals and means [Rokeach 1973: 29]. Two lists with eighteen values in each were then compiled; he retained in these lists only positive values, a single value for each group of synonyms, values maximally different from others, and values judged to be meaningful in all cultures. The author felt that she could not do better than accept Rokeach's two lists although she modified these so that respondents would focus on their life in work situation rather than their total life situation. The Rokeach and "modified" Rokeach lists are shown in Tables 1 and 2.

The author was not just interested in gaining an understanding of the broader

Table 1. Life values (values associated with goals)

Rokeach model	Modified Rokeach	
A comfortable life (a prosperous life)	An absence of conflict (with other individuals/groups)	M*
An exciting life (a stimulating, active life)	A stimulating life in work (challenging, exciting)	C
A sense of accomplishment (lasting contribution)	A sense of accomplishment (feeling of making a contribution)	C
A world at peace (free of war and conflict)	A successful life in work (well-paid, status conferring)	C
A world of beauty (beauty of nature and the arts)	An attractive environment	CA
Equality (brotherhood, equal opportunity for all)	Equality of opportunity	M
Family security (taking care of loved ones)	Emotional security (support from colleagues and superiors)	CA
Freedom (independence, free choice)	Freedom (independence, free choice)	CA
Happiness (contentedness)	Happiness (contentedness)	CA
Inner harmony (freedom from inner conflict)	Psychological health (freedom from anxiety)	CA
Mature love (sexual and spiritual intimacy)	Good relationships (cooperation, help)	C
National security (protection from attack)		
Pleasure (an enjoyable, leisurely life)	Pleasure (an enjoyable life in work)	CA
Salvation (saved, eternal life)		
Self respect (self esteem)	Self respect (self esteem)	CA
Social recognition (respect, admiration)	Social recognition (respect, admiration)	
True friendship (close companionship)	Friendship (close companionship)	CA
Wisdom (a mature understanding of life)	Wisdom (a mature understanding of life)	C

* C indicates cognitive goals related to achieving knowledge, understanding, intellectual stimulus, efficiency; CA indicates cathectic goals related to affective feelings of well-being, pleasure; M indicates moral goals related to moral, "right and wrong" choices.

65

Table 2. Life values (values associated with the means for achieving goals)

Rokeach model	Modified Rokeach*	
Ambitious (hard-working, aspiring)	Ambitious (hard-working)	C°
Broadminded (open-minded)		
Capable (competent, effective)	Capable (competent, effective)	C
Cheerful (lighthearted, joyful)	Cheerful (lighthearted, gay)	CA
Clean (neat, tidy)	Clean (neat, tidy)	CA
Courageous (standing up for your beliefs)	Courageous (standing up for one's beliefs)	M
Forgiving (willing to pardon others)	Forgiving (willing to accept the deficiencies of others)	M
Helpful (working for the welfare of others)	Helpful	M
Honest (sincere, truthful)	Honest (sincere, truthful)	M
Imaginative (daring, creative)	Creative (imaginative)	CA
Independent (self-reliant, self-sufficient)	Independent (self-reliant)	C
Intellectual (intelligent, reflective)	Open-minded (receptive to new ideas)	C
Logical (consistent, rational)	Logical (consistent, rational)	C
Loving (affectionate, tender)		
Obedient (dutiful, respectful)	Obedient (dutiful, respectful)	M
Polite (courteous, well-mannered)	Polite (courteous, well-mannered)	CA
Responsible (dependable, reliable)	Responsible (dependable, reliable)	C
Self-controlled (restrained, self-disciplined)	Self-controlled (restrained, self-disciplined)	C

* The "modified Rokeach" was put into alphabetical order for use in the research situation.
° C indicates a cognitive means: the most efficient way of achieving a goal; CA indicates a cathectic means: the most pleasant way of achieving a goal (Parsons would use the term "expressive". In order to avoid confusion, "cathectic" will be used from now on.); M indicates moral means: the correct or moral way of achieving a goal.

values of the systems designers, and in comparing these with the values of user and top management, she also wished to know if these values were predominantly what Parsons has called "cognitive"[3] goals. That is goals associated with achieving knowledge, rational understanding, intellectual stimulus and the pursuit of efficiency; "cathectic"[4] or appreciative goals related to the attainment of a state of well-being or pleasure, or moral goals associated with achieving something viewed as morally "right". Similarly, she wished to know if the values associated with "means" rather than "ends" were cognitive in the sense that they were related to the most efficient way of reaching a goal; cathectic (involving expressive action) the most pleasant way, or moral, the most "correct" or morally right way. Parsons' cognitive, cathectic and moral categories were therefore associated with Rokeach's two lists. This was not easy, as some goals could be defined in different ways and placed in more than one category, but on the whole the Rokeach values fitted the Parsonian value classification fairly readily.

Each respondent was given the two lists of values with the values in alphabetical order. Each value was printed on a removable gummed label so that it could be transferred easily to a blank column on the opposite side of the page, and also moved about if the respondent changed his mind. Systems designers and managers were asked to rank these values according to their importance as guiding principles in their own lives and behaviour. The hypothesis here was that the systems designers, in common with other managerial groups, would have predominantly cognitive rather than cathectic or moral values.

This hypothesis was found to be correct; although the goal-directed value list contained only 5 cognitive values, compared with 9 cathectic and 2 moral, the upper quartile of the responses was predominantly in terms of cognitive values. 11 systems designers ranked the goals list which meant that there were 44 responses in the upper quartile; of these 31 were cognitive.[5] In the lower quartile representing those values on which systems designers placed least importance, only 5 responses were in the cognitive category. A similar pattern was found with the "means" list of values in which 7 were cognitive, 4 cathectic or expressive and 5 moral. 12 systems designers answered this producing 48 responses in the upper quartile. Of these 35 were cognitive; only 2 cognitive responses were found in the lower quartile.

---

3. Cognitive goals relate to getting things done rationally and efficiently.
4. Cathectic goals relate to what is pleasurable or painful.
5. The reason for a research population of only eleven was that these were the systems designers primarily responsible for the design of the systems in the case study organizations. The intention was to relate their values to their design philosophy and practice.

Cognitive goals that were referred to *most* frequently were:

—A stimulating life in work (challenging, exciting): 11 of the responses in the upper quartile[6] (out of a total of 44 responses);
—A sense of accomplishment (feeling of making a contribution, doing a good job): 9 responses.

Goals that were referred to most frequently in the lower quartile, and therefore seen as *least* important were:

—Equality of opportunity: 7 responses;
—Friendship (close companionship): 6 responses;
—Social recognition (respect, admiration): 6 responses.

In the "means for achieving goals" list the behaviour that was referred to *most* frequently came into the cognitive category and was being:

—Capable (competent, effective): 9 of the responses in the upper quartile (48 responses);
—Responsible (dependable, reliable): 9 responses.

Creativity in the sense of being imaginative, which we have defined as cathectic rather than cognitive, although it perhaps fits both categories, received 7 responses.

Behaviour referred to most frequently in the lower quartile and therefore seen as *least* important was being:

—Obedient (dutiful, respectful): 11 responses;
—Polite (courteous, well-mannered): 10 responses;
—Clean (neat, tidy): 9 responses.

The picture that emerged from this analysis, of cognitively oriented people focussed on goals and means associated with intellectual and rational behaviour was also true of the senior and departmental managers in charge of the user areas in the research situations. 12 managers completed the ranking which meant that there were 48 responses in the upper quartile. In the goals list, 30 of these were cognitive; in the means list, 36 were cognitive. The managers' first and second choices, like those of the systems designers, were "a sense of accomplishment" and "a stimulating life in work". The pattern for the systems designers, departmental managers and senior managers was as shown in Tables 3 and 4. The only differences in the preferences are that systems designers approved of "creative" ways of doing things, managers of "logical" ways and senior managers of "honest" ways.

---

6. The quartile has been taken as the unit of analysis as it is thought to be more meaningful than taking first, second choices, etc. People completing the rankings often said they would like to give items equal weight.

Table 3. Most desired goals

Systems designers N = 11 (44 responses)	User managers N = 12 (48 responses)	Senior managers N = 6 (24 responses)
A stimulating life in work (11)	A sense of accomplishment (10)	A sense of accomplishment (6)
A sense of accomplishment (9)	A stimulating life in work (9)	A stimulating life in work (5)

Other goals had only small numbers of supporters.

Table 4. Most approved of means

Systems designers N = 12° (48 responses)	User managers N = 12 (48 responses)	Senior managers N = 6 (24 responses)

Being:
  Capable (9)
  Responsible (9)
  Creative (7)
	Being:
	  Capable (9)
	  Responsible (8)
	  Logical (6)
		Being:
		  Honest (5)
		  Capable (4)
		  Responsible (4)

° One systems designer was only prepared to rank the "means" list claiming that he no longer had any goals in the company.

Table 5. Most desired goals

Systems designers		All managers	
A stimulating life in work	C	A stimulating life in work	C
A sense of accomplishment	C	A sense of accomplishment	C
Pleasure	CA	A successful life in work	C
A successful life in work	C	Self-respect	CA
Good relationships (cooperation)	C	Wisdom	C
Equality of opportunity	M	Good relationships (cooperation)	C
Self-respect	CA	An absence of conflict	M
Emotional security	CA	Freedom	CA
Freedom	CA	Pleasure	CA
Social recognition	CA	Happiness	CA
Psychological health	CA	An attractive environment	CA
Happiness	CA	Emotional security	CA
Friendship	CA	Friendship	CA
An absence of conflict	M	Psychological health	CA
An attractive environment	CA	Equality of opportunity	M
Wisdom	C	Social recognition	CA

C refers to Cognitive goals, CA to Cathectic goals, M to Moral goals.

The systems designers' ranking of all "goal" statements is shown in the left hand column of Table 5, with the managers ranking in the right hand column (both user and senior managers are included here). These rankings are derived from those variables which received the largest number of responses in each choice category.[7] Occasionally the same variable appeared at different choice levels. When this happened it was allotted to the position in which it received the most responses.

Some interesting differences emerge from these rankings. Our systems designers placed a lower priority on "wisdom" as an important goal, and on "an absence of conflict". They placed higher priority on "equality of opportunity" and on "pleasure". Although too much trust must not be placed in the ranking positions as both systems designers and managers said it was not easy to make this kind of choice, nevertheless these large differences may have some significance. There is, however, considerable agreement within each group in the upper and lower quartiles. The most interesting observation is again the importance of cognitive goals to both groups (Table 6).

Table 6. Most approved means for achieving goals

Systems designers		Managers	
Being:			
Capable	C	Capable	C
Ambitious	C	Responsible	C
Independent	C	Honest	M
Responsible	C	Independent	C
Creative	CA	Logical	C
Logical	C	Ambitious	C
Open-minded	C	Courageous	M
Self-controlled	C	Self-controlled	C
Honest	M	Forgiving	M
Helpful	M	Creative	CA
Forgiving	M	Open-minded	C
Courageous	M	Obedient	M
Polite	CA	Polite	CA
Obedient	M	Cheerful	CA
Clean	CA	Helpful	M
Cheerful	CA	Clean	CA

C refers to Cognitive means, CA to Cathectic means, M to Moral means.

---

7. Rokeach bases his rankings on medians but the numbers here are too small for either medians or means to show up differences.

These two sets of rankings do not diverge greatly from each other and again the fact of greatest interest is that cognitive or rational means are given highest priority. What does this analysis tell us? It suggests that we have here two groups of people who are similar in that they are achievement-oriented and looking for intellectual challenge and stimulus in work. They value efficiency and the capability and responsibility that lead to this. At this stage we can distinguish few major differences between the values of systems designers and the values of managers using their systems.

4.2. ORGANIZATIONAL VALUES

After the systems designers and user managers had described their values in relation to their own lives in work, they were asked to indicate on a scale *the values of their firms, bank or civil service departments, as they saw them,* and also to show what they would *ideally like* the values of these organizations to be. The basis of this analysis was the Parsonian value framework described in chapter 3. Systems designers and managers were now given two sets of statements, representing the extremes of a scale graded 1-7 and asked to mark where their own opinions lay on this scale. All clerks were given the same statements and scale and Figures 1-4 show the means and modes of the answers of all three groups. Modes are shown as well as means, as we are more interested in responses at the extremes of the axis than in central tendencies. Because we are dealing with aggregate data derived from a number of very different situations, at this stage of the analysis we are less interested in where each mean or mode appears on the scale than in the extent to which there are similarities or divergences in the views of the systems designers, managers or clerks.

Familiarity with the protestant ethic values of much of British industry suggests that we can hypothesize that all our research situations will be seen as wanting shared values, discipline, standardized methods and procedures, efficiency and high production, and structured jobs which do not provide employees with much discretion. We can also hypothesize that many staff will want organizational values to be more flexible, our argument here being that the role expectations of employees are changing at a faster rate than organizations are altering their structures. Any change in role expectations, it is anticipated, will be towards more self-management, control and discretion. Influenced by the earlier stereotypes of the technical specialist we can also hypothesize that the value profiles of the systems designers will not be the same as those of the managers and clerks. The results are displayed in Figures 1 and 2. It can be seen that on the perceived organizational values scale (Figure 1), systems designers and managers have very similar profiles when the mean is used as the measure; they diverge only on the fourth category with the managers seeing the firm as placing more emphasis on personal qualities. In contrast the clerks see their organizations as being further to the left of the axis; particularly in relation to disciplined employees and efficiency. All three groups fulfil our expectations that they will see organizational values as focussed on discipline, efficiency and structured jobs; with shared values as the means for getting

Figure 1. Perceived organizational values (means)

Please tick where you think this firm, bank, civil service department fits on the scale below

It has a strong belief in the importance of shared values – that employees should agree with its objectives and the way it carries out its operations

It likes to have disciplined employees who will put the firm's/office's interests first and willingly accept orders and instructions

It prefers to use standardized methods and procedures whenever possible as it believes that this assists efficiency in this kind of firm/office

It places a great deal of emphasis on efficiency and high production and less on personal qualities such as friendliness, trustworthiness, cooperation etc.

It feels the need to organize work activities into tightly structured jobs which are clearly defined and do not permit a great deal of individual discretion

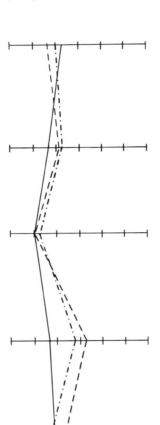

A multiplicity of different values exists and is tolerated here. It is quite acceptable for employees to hold very different views on what the firm should be doing and how it should be doing it

It does not believe in very strict discipline and tries to provide a situation in which people can pursue their own interests

It likes employees to work out their own methods for doing things and to use their own judgement when taking decisions

It places a great deal of emphasis on personal qualities such as friendliness, trustworthiness, cooperation etc. and less on efficiency and high production

It tries to organize work in such a way that employees have loosely defined and structured jobs which permit a great deal of individual discretion

———————— N = 436 All clerks
— · — · — · — N = 13 Systems designers
— — — — — — N = 18 All managers

72

Figure 2. Desired organizational values (means)

Please tick where you would like this firm, bank, civil service department to fit on the scale below.

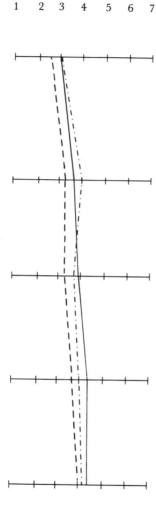

It should have a strong belief in the importance of shared values – that employees should agree with its objectives and the way it carries out its operations

A multiplicity of different values should exist and be tolerated here. It should be quite acceptable for employees to hold very different views on what the firm should be doing and how it should be doing it

It should try to create disciplined employees who will put the firm's/office's interests first and willingly accept orders and instructions

It should have an easy-going kind of discipline and try to provide a situation in which people can pursue their own interests

It should try to use standardized methods and procedures whenever possible as this would assist efficiency in this kind of firm/office

It should encourage employees to work out their own methods for doing things and use their own judgement when taking decisions

It should place a great deal of emphasis on efficiency and high production and less on personal qualities such as friendliness, trustworthiness, co-operation etc.

It should place a great deal of emphasis on personal qualities such as friendliness, trustworthiness, cooperation etc. and less on efficiency and high production

It should try to organize work activities into tightly structured jobs which are clearly defined and do not permit a great deal of individual discretion

It should try to organize in such a way that employees have loosely defined and structured jobs which permit a great deal of individual discretion

————— N = 436 All clerks
– · – · – · – N = 13 Systems designers
– – – – – – N = 18 All managers

73

employees to accept these requirements. On the desired organizational values (Figure 2), there is agreement that shared values are something that should be encouraged, but some divergence on the other value positions. The systems designers are further to the right than the other groups in wanting an easy-going type of discipline, perhaps reflecting their own management services culture, and clerks are everywhere further towards the mid point of the scale than managers. Similarly, all three groups move away from the "standardized methods" category in the "desired values" chart, although only towards the mid point. The clerks want more emphasis on personal qualities and friendliness, the managers less. All three groups favour less emphasis on tightly structured jobs. It is interesting that although all three groups would like their organizations to be slightly more liberal and flexible, they do not appear to want any major change and the "desired values" profiles of the systems designers and clerks hover around the mid point of the scale. The "mode" diagrams (Figures 3 and 4) are more informative and show the differences between the groups more clearly. Systems designers now fall between the managers and the clerks but, whereas the profiles for "perceived" and "desired" values change little for the systems designers and managers, the modes of the clerks on the three variables, discipline, standardized methods and tightly structured jobs, move from the extreme left to the centre of the scale. There is also a considerable discrepancy between the views of the managers and the systems designers on the structuring of work. The managers believe that jobs should be more tightly structured, while the systems designers move towards the "loosely defined and structured jobs" statement, one group moving to point 5 on the scale. Managers and systems designers show more agreement with each other than they do with the clerks. So far, our first two hypotheses are supported; but not the third, that systems designers will have values different from those of the other two groups.

Many of the systems designers explained why they chose the left hand rather than the right hand side of the scale for "desired organizational values". Government systems designers pointed out that their organizations were subject to legal constraints. "Efficiency in the sense of accuracy of payment according to the law is essential", said a systems designer in one government department, and this kind of comment was echoed by systems designers in the other two. "Because of legal constraints a too loose system of discipline or too many private methods of working are not possible", was a typical comment. Another was, "Because we are a government department we have to place great importance on accounting, security and prevention of fraud. All operations must therefore be strictly controlled, although there is consultation at all levels on how we achieve our objectives". The comments of the systems designers in the two firms and the bank were that although their firms still believed in standardization and structured jobs, things were relaxing.

The next set of questions which are related to the organizational models of the different groups ask for opinions on *the best form of organizational structure for clerical staff in general*. This is related to Parsons' "self-orientation – collectivity-orientation" pattern variable which distinguishes between role expectations that

Figure 3. Perceived organizational values (modes)

Please place a tick where you think this firm, bank, civil service department fits on the scale below.

It has a strong belief in the importance of shared values – that employees should agree with its objectives and the way it carries out its operations	A multiplicity of different values exists and is tolerated here. It is quite acceptable for employees to hold very different views on what the firm should be doing and how it should be doing it
It likes to have disciplined employees who will put the firm's/office's interests first and willingly accept orders and instructions	It does not believe in very strict discipline and tries to provide a situation in which people can pursue their own interests
It prefers to use standardized methods and procedures whenever possible as it believes that this assists efficiency in this kind of firm/office	It likes employees to work out their own methods for doing things and to use their own judgement when taking decisions
It places a great deal of emphasis on efficiency and high production and less on personal qualities such as friendliness, trustworthiness, cooperation etc.	It places a great deal of emphasis on personal qualities such as friendliness, trustworthiness, cooperation etc. and less on efficiency and high production
It feels the need to organize work activities into tightly structured jobs which are clearly defined and do not permit a great deal of individual discretion	It tries to organize work in such a way that employees have loosely defined and structured jobs which permit a great deal of individual discretion

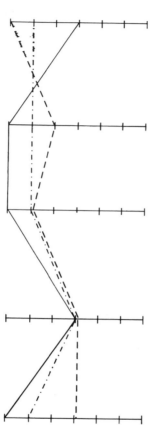

——————— N = 436   All clerks
– · – · – · – N = 13   Systems designers
– – – – – – N = 18   All managers

75

Figure 4. Desired organizational values (modes)

Please tick where you would like this firm, bank, civil service department to fit on the scale below.

It should have a strong belief in the importance of shared values – that employees should agree with its objectives and the way it carries out its operations

A multiplicity of different values should exist and be tolerated here. It should be quite acceptable for employees to hold very different views on what the firm should be doing and how it should be doing it

It should try to create disciplined employees who will put the firm's/office's interests first and willingly accept orders and instructions

It should have an easy-going kind of discipline and try to provide a situation in which people can pursue their own interests

It should try to use standardized methods and procedures whenever possible as this would assist efficiency in this kind of firm/office

It should encourage employees to work out their own methods for doing things and use their own judgement when taking decisions

It should place a great deal of emphasis on efficiency and high production and less on personal qualities such as friendliness, trustworthiness, co-operation etc.

It should place a great deal of emphasis on personal qualities such as friendliness, trustworthiness, cooperation etc. and less on efficiency and high production

It should try to organize work activities into tightly structured jobs which are clearly defined and do not permit a great deal of individual discretion

It should try to organize in such a way that employees have loosely defined and structured jobs which permit a great deal of individual discretion

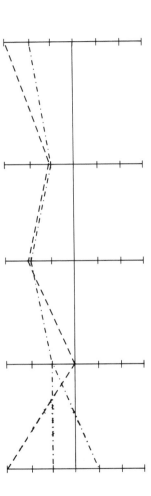

——————— N = 436   All clerks
– · – · – · – N = 13   Systems designers
– – – – – N = 18   All managers

76

an individual or group will want to pursue their own particular interests and the role expectation that they will prefer to conform with wider organizational interests. Today we would see one aspect of this difference as flexible self-management versus more rigid management imposed by an organization through its supervisory system. These questions had formed part of the pilot study carried out in Britain and Sweden [Hedberg and Mumford 1975] (see Figures 7, 8 and 12). The pilot results which were presented to an international conference, showed that neither British nor Swedish systems designers were very far over towards the self-management or "Theory Y"-side of the axis, although the Swedes were more theory Y-oriented than the British. These data were widely reported and were seen as supporting the view that systems designers were illiberal, did not understand user needs and interests and were more interested in facilitating organizational values than individual or task group values. As the author was writing this chapter she came across an article in the current *Computer Weekly* stating that the work of Mumford and Hedberg had "proved" that systems designers are theory X; a "fact" which the writer of the article deplored. In this way invalid evidence becomes part of people's mental models and takes on the guise of truth [Phillips 1973: 14]. We can now check whether the values of the systems designers in the present research are theory X or theory Y and, more important, establish whether they are similar to, or diverge from the values of the managers and clerks. As we are now looking at values related to departmental structure and organization, we will separate the views of departmental (user) managers from those of top managers.

The statistical means are not very informative as they pull the data to the centre of the scale but we can see that the profiles of the four groups resemble each other, swinging over to the left on "targets should be set by supervision" and to the right on "everyone should have access to all information". The modes provide more information and we find that the systems designers' profiles resemble those of the user manager. The double modes show a split in the systems designer group on whether jobs should be defined by O&M, management services, etc., or left to the group and individual doing the job. Clerks and user managers are over to the right hand side of the scale on "everyone should have access to all information" and senior managers and clerks are over to the right on "decisions should be arrived at through group discussions involving all employees". The systems designers do not come out as markedly different from other groups, however, and on "delegation of authority and responsibility to those doing the job" are further to the right, with management, than the clerks themselves.

How can we interpret these data? The first point to make is that reporting the pilot study data to a conference without having comparable data from user groups, may have done systems designers an injustice. It is likely, although Mumford and Hedberg did not know this, that their values did not diverge greatly from the values of the users of their systems.[8] The second point is that so far in

---

8. The pilot study did provide evidence that the systems designers had an unflattering view of users. 81% of the British comments referred to shortcomings, suggesting that they

this analysis we have no evidence to suggest that the values of systems designers reflect the technical anti-human orientation of the popular stereotype. On the contrary, in our research situations, their values conform pretty closely to those of the management and clerical groups. Figures 7, 8, 9 and 12 show the profiles of our systems designers compared with the pilot study groups and with American engineers and systems designers. Perhaps the most interesting thing here is the similarity of all the profiles, with the exception of the Swedish modes. There seems to be a consensus on how these questions should be answered. However, normative responses do not necessarily imply normative behaviour and we may find that behaviour diverges from attitudes when we examine our case studies.

A third and last value scale asked systems designers, managers and clerks to indicate *how they saw the majority of staff in the departments using the computer systems.* The clerks were now commenting on themselves and their own characteristics. Again the statistical means produced similar profiles; with the clerks and senior managers further over to the Y-side of the axis than the systems designers and user managers. The modes showed a wider spread of responses. Systems designers and the managers of user departments were on the X-side of the axis for three and four of the seven variables, respectively, and favoured "targets being set by supervision", and "output and quality standards being monitored by supervision". Both groups believed that the clerks worked best on jobs with a short task cycle and departmental managers also thought that the clerks "liked to be told what to do next and how to do it". The modes of the senior managers and clerks were on the Y-side of the axis. The clerks claimed that they, as a group responded well to "varied, challenging work, requiring knowledge and skill"; they also indicated that "they liked, and were competent to use initiative and take decisions". Top management and the systems designers thought that the clerks regarded opportunities for social contact at work as important, but the clerks themselves were at the mid point of the scale on this variable.

Here, once again, we have a value scale on which the views of user management and the systems designers are close, with the managers being more theory-X than the systems designers. Figure 12 compares the means of the British and Swedish systems designers in the pilot study with our present group and with the American engineers and systems designers. Again the profiles are close with the Swedish group being most Y-oriented on two variables, the ability of Swedish clerks to undertake varied work and take responsibility for decisions; and most X-oriented on the needs of the clerks to be told what to do and how to do it.

Only four questions are shown in Figure 12 as the wording of some of the questions was changed after the American and pilot studies had been completed.

---

were "institutionalized" (13%), resistant to change (29%), and limited in outlook (39%). 54% of the Swedish comments were similar.

78

Figure 5. Organizational models (means)

Please tick on the scale below what *you* believe to be the best form of department structure for clerical staff in general.*

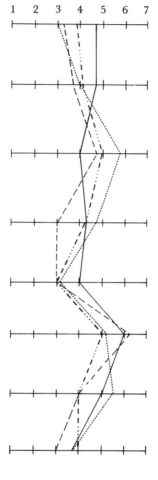

	1 2 3 4 5 6 7	
Jobs should be clearly defined, structured and stable		Jobs should be flexible and permit group problem solving
There should be a clear hierarchy of authority with the man at the top carrying ultimate responsibility for all aspects of work		There should be a delegation of authority and responsibility to those doing the job regardless of formal title and status
The most important motivators should be financial, e.g., high earnings and cash bonuses		The most important motivators should be non-financial, e.g., work challenge, opportunity for team work
Jobs should be carefully defined by O&M department, management services or supervision and adhered to		The development of job methods should be left to the group and individual doing the job
Targets should be set by supervision and monitored by supervision		Targets should be left to the employee groups to set and monitor
Groups and individuals should be given the specific information they need to do the job but no more		Everyone should have access to *all* information which they regard as relevant to their work
Decisions on what is to be done and how it is to be done should be left entirely to management		Decisions should be arrived at through group discussions involving all employees
There should be close supervision, tight controls and well maintained discipline		There should be loose supervision, few controls and a reliance on employee self-discipline

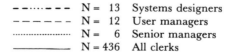

─ ─·····─ ─ ─	N =	13	Systems designers
─ ─ ─ ─ ─ ─	N =	12	User managers
····················	N =	6	Senior managers
────────	N = 436		All clerks

* These questions were randomized in the original questionnaire. They are presented here with theory-X statements down one side and theory-Y down the other.

Figure 6. Organizational models (modes)

Please tick on the scale below what *you* believe to be the best form of department structure for clerical staff in general.*

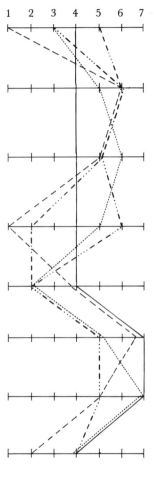

Jobs should be clearly defined, structured and stable	Jobs should be flexible and permit group problem solving
There should be a clear hierarchy of authority with the man at the top carrying ultimate responsibility for all aspects of work	There should be a delegation of authority and responsibility to those doing the job regardless of formal title and status
The most important motivators should be financial, e.g., high earnings and cash bonuses	The most important motivators should be non-financial, e.g., work challenge, opportunity for team work
Jobs should be carefully defined by O&M department, management services or supervision and adhered to	The development of job methods should be left to the group and individual doing the job
Targets should be set by supervision and monitored by supervision	Targets should be left to the employee groups to set and monitor
Groups and individuals should be given the specific information they need to do the job but no more	Everyone should have access to *all* information which they regard as relevant to their work
Decisions on what is to be done and how it is to be done should be left entirely to management	Decisions should be arrived at through group discussions involving all employees
There should be close supervision, tight controls and well maintained discipline	There should be loose supervision, few controls and a reliance on employee self-discipline

- – .... – – N = 13 Systems designers
- – – – – N = 12 User managers
.... N = 6 Senior managers
———— N = 436 All clerks

* These questions were randomized in the original questionnaire. They are presented here with theory-X statements down one side and theory-Y down the other.

80

Figure 7. Organizational models (means)

Results of 1975 study of values compared with the results of the study described in this book.

Views of British and Swedish systems designers on the best form of department structure for non-specialist, non-management staff, compared with the views of systems designers in this study.

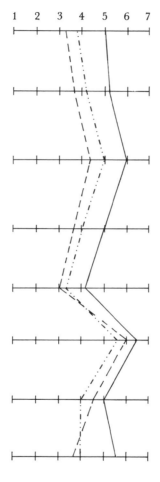

Jobs should be clearly defined, structured and stable	Jobs should be flexible and permit group problem solving
There should be a clear hierarchy of authority with the man at the top carrying ultimate responsibility for all aspects of work	There should be a delegation of authority and responsibility to those doing the job regardless of formal title and status
The most important motivators should be financial, e.g., high earnings and cash bonuses	The most important motivators should be non-financial, e.g., work challenge, opportunity for team work
Jobs should be carefully defined by O&M department, management services or supervision and adhered to	The development of job methods should be left to the group and individual doing the job
Targets should be set by supervision and monitored by supervision	Targets should be left to the employee groups to set and monitor
Groups and individuals should be given the specific information they need to do the job but no more	Everyone should have access to *all* information which they regard as relevant to their work
Decisions on what is to be done and how it is to be done should be left entirely to management	Decisions should be arrived at through group discussions involving all employees
There should be close supervision, tight controls and well maintained discipline	There should be loose supervision, few controls and a reliance on employee self-discipline

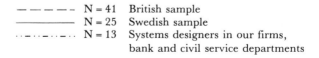

– – – – – N = 41 British sample
————— N = 25 Swedish sample
· · — · · — · · N = 13 Systems designers in our firms, bank and civil service departments

81

Figure 8. Organizational models (modes)
Results of 1975 study of values compared with the results of the study described in this book. Views of British and Swedish systems designers on the best form of department structure for non-specialist, non-management staff compared with the views of systems designers in this study.

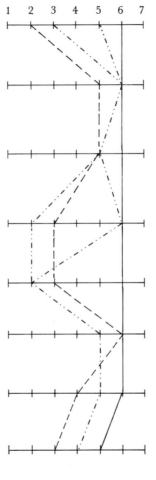

Jobs should be clearly defined, structured and stable	Jobs should be flexible and permit group problem solving
There should be a clear hierarchy of authority with the man at the top carrying ultimate responsibility for all aspects of work	There should be a delegation of authority and responsibility to those doing the job regardless of formal title and status
The most important motivators should be financial, e.g., high earnings and cash bonuses	The most important motivators should be non-financial, e.g., work challenge, opportunity for team work
Jobs should be carefully defined by O&M department, management services or supervision and adhered to	The development of job methods should be left to the group and individual doing the job
Targets should be set by supervision and monitored by supervision	Targets should be left to the employee groups to set and monitor
Groups and individuals should be given the specific information they need to do the job but no more	Everyone should have access to *all* information which they regard as relevant to their work
Decisions on what is to be done and how it is to be done should be left entirely to management	Decisions should be arrived at through group discussions involving all employees
There should be close supervision, tight controls and well maintained discipline	There should be loose supervision, few controls and a reliance on employee self-discipline

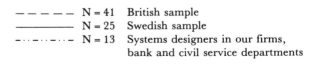

- – – – – N = 41 British sample
- ———— N = 25 Swedish sample
- – · · – · · – · · N = 13 Systems designers in our firms, bank and civil service departments

Figure 9. Organizational models (means)

Data from 1975 study of values of U.S. systems designers and engineers compared with the values of systems designers in the study described in this book.

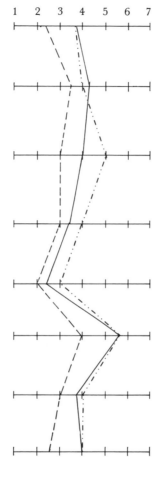

Jobs should be clearly defined, structured and stable	Jobs should be flexible and permit group problem solving
There should be a clear hierarchy of authority with the man at the top carrying ultimate responsibility for all aspects of work	There should be a delegation of authority and responsibility to those doing the job regardless of formal title and status
The most important motivators should be financial, e.g., high earnings and cash bonuses	The most important motivators should be non-financial, e.g., work challenge, opportunity for team work
Jobs should be carefully defined by O&M department, management services or supervision and adhered to	The development of job methods should be left to the group and individual doing the job
Targets should be set by supervision and monitored by supervision	Targets should be left to the employee groups to set and monitor
Groups and individuals should be given the specific information they need to do the job but no more	Everyone should have access to *all* information which they regard as relevant to their work
Decisions on what is to be done and how it is to be done should be left entirely to management	Decisions should be arrived at through group discussions involving all employees
There should be close supervision, tight controls and well maintained discipline	There should be loose supervision, few controls and a reliance on employee self-discipline

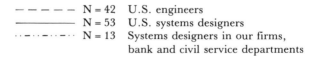

----- N = 42   U.S. engineers
————— N = 53   U.S. systems designers
·— ··— ··— ·· N = 13   Systems designers in our firms, bank and civil service departments

Source: James C. Taylor, "A Report of Preliminary Findings from the 1976 'Work Organization Study' Pilot Study" (Centre for Quality of Working Life, UCLA Reprints, 1976).

Figure 10. Models of man
Perceived characteristics of the majority of clerks in this department (means)

Please tick on the scale below how you see the general characteristics of the majority of staff in this department.

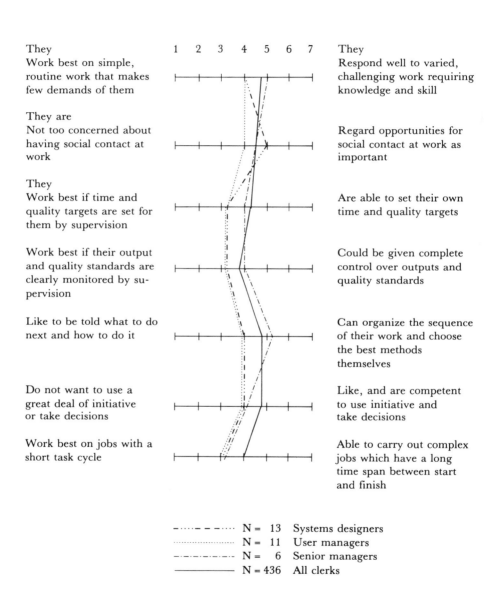

They Work best on simple, routine work that makes few demands of them	They Respond well to varied, challenging work requiring knowledge and skill
They are Not too concerned about having social contact at work	Regard opportunities for social contact at work as important
They Work best if time and quality targets are set for them by supervision	Are able to set their own time and quality targets
Work best if their output and quality standards are clearly monitored by supervision	Could be given complete control over outputs and quality standards
Like to be told what to do next and how to do it	Can organize the sequence of their work and choose the best methods themselves
Do not want to use a great deal of initiative or take decisions	Like, and are competent to use initiative and take decisions
Work best on jobs with a short task cycle	Able to carry out complex jobs which have a long time span between start and finish

----- N = 13   Systems designers
·········· N = 11   User managers
-·-·-·-·- N = 6   Senior managers
————— N = 436   All clerks

84

Figure 11. Models of man
Perceived characteristics of the majority of clerks in this department (modes)

Please tick on the scale below how you see the general characteristics of the majority of staff
in this department.

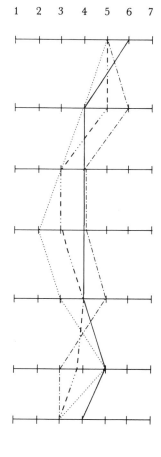

They	1 2 3 4 5 6 7	They
Work best on simple, routine work that makes few demands of them		Respond well to varied, challenging work requiring knowledge and skill
They are Not too concerned about having social contact at work		Regard opportunities for social contact at work as important
They Work best if time and quality targets are set for them by supervision		Are able to set their own time and quality targets
Work best if their output and quality standards are clearly monitored by supervision		Could be given complete control over outputs and quality standards
Like to be told what to do next and how to do it		Can organize the sequence of their work and choose the best methods themselves
Do not want to use a great deal of initiative or take decisions		Like, and are competent to use initiative and take decisions
Work best on jobs with a short task cycle		Able to carry out complex jobs which have a long time span between start and finish

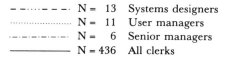

```
– –·····– – –   N = 13   Systems designers
················   N = 11   User managers
·–·–·–·–·–   N =  6   Senior managers
————————   N = 436   All clerks
```

85

Figure 12. Models of man (means)

Views of British and Swedish systems designers and American production engineers and systems designers of the typical employee (non-supervisory and non-specialist) for whom they design work systems; compared with views of systems designers in this study of clerks in their user departments.*

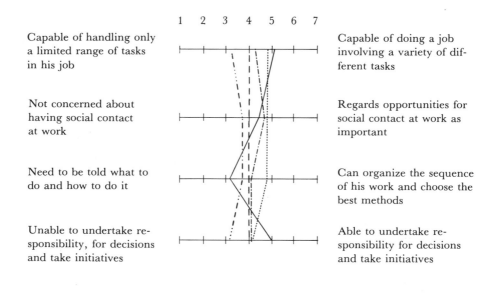

Capable of handling only a limited range of tasks in his job		Capable of doing a job involving a variety of different tasks
Not concerned about having social contact at work		Regards opportunities for social contact at work as important
Need to be told what to do and how to do it		Can organize the sequence of his work and choose the best methods
Unable to undertake responsibility, for decisions and take initiatives		Able to undertake responsibility for decisions and take initiatives

———————— N = 25 Swedish study (pilot)
- - - - - - N = 41 British study (pilot)
- ··· - - - ···- N = 42 American Engineers
················· N = 53 American systems designers
—·—·—·—·—·— N = 13 British systems designers in this study

* A number of statements were reworded for the present research. Only those statements in which the wording is identical or very close are shown here.

4.3. CONCLUSIONS

The intention of this chapter has been to obtain a view of the values and mental models of systems designers relating to their employing organizations, to the best form of departmental structure for clerks and to the needs of clerks, and to test the hypothesis that the values of systems designers are different from those of other non-technical groups. The reasons for generating this hypothesis were that a stereotype of the technically oriented systems designer with little sympathy for human needs has been prevalent in the contributions of humanists and social scientists to the literature on systems design. Also previous research had shown that groups of systems designers in both Britain and Sweden held organizational models and models of man which were not, in McGregor's terms, very Y-oriented. Two hypotheses were derived from this earlier research. The first, that the values of systems designers would influence the nature of the systems they built. The second, that the values of staff in departments using computer systems would be more Y-oriented than those of the designers of the systems; therefore there would be a poor fit between job needs and the ability to realize these needs. A theory X-design orientation would lead to theory X-work systems with management-imposed controls and tightly structured jobs.

It seemed important to test the hypothesis concerning values before embarking on a detailed analysis of the case study material. Otherwise analysis of the data might incorporate erroneous assumptions about the philosophy and attitudes of the systems designers. The results of this test have proved fruitful. It seems that the values of the systems designers in our firms, bank and government departments are similar to those of the user managers, with a tendency to be more accepting of theory Y-flexibility. This is true of the scales for "desired organizational values", "organizational models" of the best form of departmental structure, and the models of man – the "perceived characteristics of clerks in the user departments". Systems designers values therefore appear to reflect those of middle management.

Systems designers often argue that they would like to take a more humanistic approach to systems design but are prevented from doing so by the values of top management, whom they see as principally interested in efficiency. However, the senior managers in our organizations do not, so far, fit this description. Their value positions on the organizational model and models of man scales tend to be to the right of the systems designers and user managers, and even to the right of the clerks on a number of variables. Systems designers may be assuming attitudes in top management which do not always exist and it will be interesting to check this in the case studies.

The principal divergence of values in our data is not between the systems designers and other groups, but between user managers/systems designers and the clerks. On all three value scales clerks are more Y-oriented; for example, wanting to set their own targets and take their own decisions. They also regard themselves as responding well to challenging work, able to assume control of output and quality standards and able to carry out complex jobs with a long time span between start and finish. On the Parsonian pattern variables they, as a group, have the biggest gap between "perceived" and "desired" values. They would like their organizations to place less emphasis on collective interests and more on personal and group interests (Parsons' collectivity-orientation versus self-orientation); less emphasis on standardized methods (universalism versus particularism), and on tightly organized jobs (specificity versus diffuseness). Even so their desires are modest and only on the mid point of the value scale.

To sum up, in this study we are concerned with a group of systems designers whose values closely reflect those of the middle managers who use their systems. This changes the focus of the study from an emphasis on why systems designers' values diverge from those of other groups, to an examination of organizational values and how these influence both the design of a computer system and the extent to which it is able to meet both organizational needs and the need dispositions and the role expectations of the clerks who use it.

5. The Research Situations, Research Methodology and Framework of Analysis of the Case Studies

The author was enabled to conduct the research for this book through the award of a personal research grant by the Social Science Research Council. This meant that she had one year in which to carry out the necessary fieldwork. The pilot project conducted with Professor Hedberg of the University of Gothenburg had demonstrated that the issue of "values" in relation to the design of computer systems was worth pursuing in more depth, and her first requirement was to identify a number of organizations which were large users of computer systems and which would be interested in collaborating in the research. In view of the restricted research coverage that can be carried out by a lone research worker, it seemed important to choose organizations which fell into two broad groups so as to enable research comparisons to be made. Within each group there should be some similarity in the principal independent variable, namely values.

The author has had a long association with the Central Computer Agency of the Civil Service department and through the kind offices of its Director, Mr. Reay Atkinson, she was introduced to three major government departments, one of which was the Inland Revenue. Meetings with senior officials of these three departments enabled specific research sites to be chosen with the Inland Revenue choice being Centre 1, at East Kilbride in Scotland. This office dealt with the tax affairs of all employees whose PAYE[1] tax was paid in Scotland by their employers.

With the aim of reducing the number of independent variables the author had decided to choose organizations which had a reputation for good employee relations and for paying a great deal of attention to the needs of their staff. In such situations it might be expected that there would be pressures on systems designers and managers to consider staff needs when designing computer systems and therefore to approach systems design problems with a similar kind of humanistic value orientation. The Civil Service seemed particularly appropriate in this respect and, in addition, it had an excellent, long established consultative system which might also influence the way computer systems were designed.

In order to secure useful research comparisons the organizations in the second group had to be unconnected with the Civil Service and yet to be seen as placing a great deal of importance on employee relations and the needs of their staff. The

---

1. Pay As You Earn.

author was fortunate in already having connections with two such organizations, Asbestos Ltd., a firm making asbestos products for the building industry and located in Trafford Park, Manchester, and an international bank in London. The systems design group at Asbestos Ltd. had for a long time been interested in the human consequences of their work and had involved the author in many discussions on this subject. She had been able to observe and be involved in the design processes associated with the introduction of an earlier batch computer system into the sales order department and during her research year she would be able to study the conversion of this to an on-line system.

The international bank made the initial approach to the author. This bank was proposing to convert a batch computer system in its foreign exchange department into a real-time system and perceived the change as providing an opportunity for improving the quality of working life of foreign exchange clerks through associating some humanistic redesign of jobs with the new computer system. It asked for the author's assistance in doing this. Before and after comparisons should be possible in this bank through obtaining the clerks' views on the batch system and information on the design processes associated with this; then observing the design of the real-time system and obtaining the clerks' reaction to this once it was implemented. The author now required a third firm to complete her sample and she approached a large chemical company, which we shall call Chemco, influenced here by its image as an excellent employer which paid a great deal of attention to the needs and interests of its employees. She was fortunate in that she found a suitable research situation in the distribution services department of that firm. This was divided into export and home sales sections. Home sales had already converted from a batch to an on-line computer system and export was about to do so, thus presenting opportunities for before and after comparison.

The author therefore had six good research situations. Three government departments in which it would be possible to compare manual procedures with computer-based work systems; and two firms and a bank in which it would be possible to make pre- and post-change comparisons.

5.1. RESEARCH METHODOLOGY

The lone social science research worker as a rule has to tailor the approach he or she uses in collecting data to the constraints of limited time and limited interviewing resources. No team of interviewers is available to make a large scale sample of face-to-face interviews a possibility. In order to spread restricted personal resources in the most advantageous way the author decided to use a combination of face-to-face interviewing, observation, informal discussion and self-completion questionnaires. Very detailed face-to-face interviews were carried out with all systems designers, user managers and senior managers and many of these generously contributed up to a day of their time in describing their systems design philosophy and practice and completing and discussing the questions and scales on their values. The clerks presented a more difficult problem as there were large numbers of these and face-to-face interviewing was impossible. The author therefore adopted a compromise research approach that seemed to work reasonably well. All the clerks were asked to complete questionnaires which were handed out and collected by the author with guarantees that no-one else in the organization would have access to these. Management would receive only aggregate results and these would also be available to the clerks.

Because it is important to relate attitudes to the real world situation, the author spent at least a week in each office, talking informally to the clerks and closely observing, recording and discussing all the different jobs in the office. In this way she was later able to associate comments about a particular job with her own knowledge and description of the job. All questionnaires, and particularly the self-completion variety, hold the danger that it is usually the researcher that interprets the results, and these interpretations are influenced by the researcher's own values and pet hypotheses; they may therefore be far removed from the truth. Whenever possible the author avoids this difficulty by encouraging groups who have completed a questionnaire to themselves examine the results and make their own interpretations. Unfortunately it was not possible to do this in the research situations described here and so any conclusions drawn from the questionnaires completed by the clerks must be treated with caution. The interviews with the systems designers and managers were also subject to bias in that respondents were describing events which had happened in the past and memory can be a notoriously unreliable research tool. All of these are the normal problems associated with social science research and make it essential that research data is treated with caution. The data will not therefore be subjected to any sophisticated statistical tests, many of which, in the author's opinion, give a spurious respectability to data that does not warrant this degree of credibility. The aim of this book is to provide greater knowledge than we have at present and with Meehan we would define knowledge as "organized experience", and the search for knowledge as a "search for patterns of organization" [1968: 15]. Any additional knowledge acquired as a result of this research will have proved its value if it enables readers to understand the influences, particularly the influence of values, that have affected the design of a number of specific computer systems; and, from the author's own value position, if this knowledge leads to better systems design with more emphasis being placed on the successful identification and meeting of human needs. It is appropriate to end this section with a quote from Llewellyn Gross.

> . . . It may therefore be unwise for most sociologists to devote their energies to the attempted construction of abstract and comprehensive theories such as those found in the more advanced sciences. Perhaps the greatest need in sociology today is for more of the modest "inference chains", "explanation sketches", and embryo theories that aim primarily at organizing selected research findings and suggesting further avenues of inquiry. [Gross 1967: 244]

5.2. FRAMEWORK OF ANALYSIS OF CASE STUDIES

Any useful and meaningful interpretation of research data requires a framework for analysis which will assist the description and understanding of the research situations. As the framework for the analysis of values in this book has been derived from the work of Talcott Parsons, it is logical to turn to him for a tool to assist the analysis of the actions and behaviour associated with the design of new computer systems in the government department, bank and firms of our study. Parsons provides such a tool through what he calls the *functional dimensions* of any system of action [Parsons 1964a: 66]. He provides a framework for classifying action, or goal-directed behaviour, in terms of the meaning of that action for the people concerned [Parsons 1964b: 330] and also its consequences for a particular system.[2] The four components of this model are "goal attainment", "adaptation", "integration" and "pattern maintenance". Parsons believes that actions related to the attainment of goals directly reflect the values of the social system in which the action takes place and so his framework fits the theme of this book. He sees adaptive behaviour as providing facilities for the attainment of goals, while integrative acts ensure that a system subjected to change either maintains its equilibrium or reverts to a state of equilibrium once the change is completed. Pattern maintenance acts are those that help sustain the new pattern of behaviour once it is adopted [Menzies 1976: 73]. This framework with its emphasis on the relation of values to action has been used successfully by other writers, in particular Sister Marie Augusta Neal, in a study of innovation in the Roman Catholic Church [Neal 1965] and Neil Smelser in a study of the Lancashire cotton industry in the nineteenth century [Smelser 1959].

To ensure the framework fits the needs of this study, we will add another category to those of Parsons. This is "goal setting" or the kinds of choices that are made when new goals are set. A study of the goal setting process in our research organizations will enable us to understand the kinds of pressures they experienced from their environments which made them respond with a decision to make a major technical change, and the kinds of values held by the decision takers which made them choose one kind of change rather than another. Another required part of the analysis relates to the researcher's need to evaluate and draw conclusions and will provide a means for estimating the success of the change processes through the extent to which there is a good fit between the needs of the organization, its managers and its clerks.[3] Our analytical framework will therefore cover the following aspects of change. The setting of goals and the means used to *attain these goals*. Mechanisms used to ensure successful *adaptation* from an old to a new situation and the *integration* of the various subsystems into a viable and ef-

---

2. Critics of Parsons suggest that he has two different definitions of the functional dimensions here, one related to action theory, the other to systems theory [Menzies 1976: 69].

3. Parsons would see goal setting as part of the goal attainment dimension and our evaluation stage as part of integration.

fectively working whole. Lastly the strategies used to ensure that new *patterns* of values and behaviour are maintained and that there is a good fit between the needs of the different groups and the role expectations of their employing organizations. Particular attention will be paid to the influence of values. For example, do the different individuals and groups involved in the change process view it in the same way, or do they approach it from different perspectives according to their own roles and values? Ozbekhan tells us that,

> perception of a situation that is problematical, namely in need of solutions or improvement or betterment, is a function of a given value system for it is in terms of such a value system that judgements can be made as to the nature of the situation. [Ozbekhan 1969: 152]

How do values influence the manner in which change is formulated and introduced and does the experience of change alter the values of the groups affected by it? Do values influence the way the new computer systems are perceived and received? Do the most important or strongest value systems remain constant throughout the change or is there a shift in values as new knowledge and experience is obtained? All of these questions are of interest in this research.

5.2.1. Goal Setting

Under this heading we shall discuss the nature of the goals that were set prior to the design of each computer system and the methods used to determine these goals. We are also interested in who was involved in the goal-setting process and the influence of personal or group values on the choice of goals.

5.2.2. Goal Attainment

Goal attainment is the kind of action, and the values that influence this, that contributes directly to the realization of the goals that have been set. In this book it will be the strategy and behaviour associated with the systems design processes. If goals have been defined in technical terms, then goal attainment will be the processes of technical systems design. Goal attainment behaviour probably occurs simultaneously with adaptive actions directed at changing an existing set of relationships to a new set of relationships and it must be stressed that goal attainment, adaptation, integration and pattern maintenance are not an ordered sequence of activities but may occur in different sequences or simultaneously. Goal attainment usually involves both cooperation and conflict and it will be important to establish how these manifest themselves in our case study situations. It also involves the use of power, legitimized through the organization's authority structure, and the manipulation of people's perception, so that the new is seen as preferable to the old [Neal 1952: 12]. Parsons sees goals as related to equilibrium. He defines a goal as a "directional change that tends to reduce the discrepancy between the needs of the system, with respect to the input-output interchange, and the conditions in the environing system that bear upon the fulfilment of such needs" [Parsons 1961: 39]. Goals therefore are not static but interact with, and are moulded by, the environments to which they relate. Here Parsons is in accord with Stafford Beer who sees rigid goal setting and planning as merely enabling an organization to tackle today's problems on the basis of yesterday's obsolete needs [Beer 1969: 398].

5.2.3. Adaptation

Adaptation is said to have taken place when a group or unit is able to interact successfully with its environment in the sense that it gets from this environment the inputs, attitudes, goods and services, which enable it to meet its needs, and is able to return to the environment those things of which it wants to dispose [Bredemeier 1962: 51]. Adaptation is the process of moving from one state of integration or equilibrium to another and the means by which this process can take place smoothly and successfully. As any major change requires the realization of a number of different goals, not all of which will be pursued at the same time, a supply of adaptation facilities is required. These will include values, attitudes, programmes and incentives that make change acceptable and understandable [Neal 1965: 12]. Optimum adaptation is rarely achieved because the adaptation process usually requires the reconciliation of different and perhaps conflicting, interests; also most change situations have limited resources [Bredemeier 1962: 53]. Adaptation in formal organizations is therefore a political and negotiating problem in which the interests of the individual are likely to be subordinated to the interest of the group, and in which the interests of the less powerful groups may be subordinated to the interests of the more powerful. In our case study situations we shall attempt to identify and describe the adaptive facilities which were used and the success of these in producing new work systems which were successful in both human and technical terms.

5.2.4. Integration

Integration is the action taken, once goals have been attained, to restore a situation to a state of equilibrium. This involves bringing the different components of tasks, technology, people and structure together into a viable and reasonably stable relationship [Leavitt 1964: 323]. Just as adaptation and goal attainment are closely related, so adaptation and integration cannot be easily separated. An adaptive problem for a subgroup may be an integrative problem for the larger group of which the subgroup forms a part [Bredemeier 1962: 54]. For example, the problem of a management services group may be one of adapting to the company of which it forms a part; the problem for the company is one of integrating its management services group with its other functional departments. Integration is normally a group phenomenon whereas adaptation can be associated with an individual or a group.

In our case study situations we would expect two kinds of changes to have taken place, both of which will require successful integration. The first will be an adjustment to the new system by those staff whose roles have not been fundamentally changed but who are now working in a new technological environment. The second will be structural changes which involve the disappearance or reorganization of existing roles, or the creation of new ones [Smelser 1959: 14]. Successful integration implies a positive relationship between the new tasks which are a product of the new technology and the needs of the clerks who have to perform these tasks.

5.2.5. Pattern Maintenance

Once a new system has been introduced, the essential function of pattern maintenance is to maintain its stability by making it acceptable to the values and norms of the groups associated with it [Neal 1965: 11]. This requires that they view the system positively and have confidence in its ability to provide the kind of environment they need to achieve job satisfaction. It also requires the establishment of processes for socializing and educating individuals as well as the provision of what Smelser calls "tension controlling mechanisms" for handling and resolving disturbances relating to values. Smelser suggests that these tension management mechanisms are latent, or concealed, and operate independently of the system's larger adjustments to change [Smelser 1959: 11]. Ozbekhan suggests that major change requires a change of values so that a new future is "willed" [Ozbekhan 1968: 96]. Influential and innovative leaders frequently act as catalysts when this happens.

5.2.6. Evaluation of the Success of our Research Situations in Terms of Goal Attainment, Adaptation, Integration and Pattern Maintenance

In addition to Parsons' action frame of reference which provides a useful tool for studying change, the researcher requires a method for evaluating the success of the change processes. One measure will be the degree of "fit" between the need and role expectations of the different groups. A second is provided by the model shown in Table 1 which considers the consequences of the change in terms of whether these were unintended or intended, or latent or manifest.

Table 1. Classification of consequences of structures

	Consequences that foster integration and adaptation	Consequences that impede integration and adaptation
Intended or recognized consequences	Manifest functions	Manifest dysfunctions
Unintended or unrecognized consequences	Latent functions	Latent dysfunctions

Source: Bredemeier [1962: 46].

Latent, or unintended consequences, have not formed part of the original structure of goals and although their results may be beneficial to the system this has been a product of chance rather than of good management. Consequences that were intended are called "manifest" and when these are dysfunctional for the system as a whole, or for a particular group, this has been recognized in advance and is likely to be offset by some other gain which can only be secured if a related disadvantage is accepted. We shall try to identify both latent and manifest functions and the kinds of consequences which have been a product of these.

5.3. CONCLUSIONS

We expect our change processes to follow the pattern defined by Smelser and set out below [Smelser 1959: 15]:

1.   Dissatisfaction with the goal achievements of an existing system, together with a recognition that an opportunity for change has presented itself and that the facilities exist to make this change.

2.   Some symptoms of disturbance when the change is first proposed. These may take the form of "unjustified" negative emotional reactions to the proposed change and "unrealistic" aspirations regarding what can be achieved.

3.   A handling of these tensions and the mobilizing of resources to get the change underway.

4.   An encouragement of the proliferation of new ideas without, at this stage, allocating responsibility for their implementation.

5.   Positive attempts to arrive at concrete specifications for the new ideas and to create institutional mechanisms for achieving them.

6.   The responsible implementation of innovation and the reception of this with favour or disfavour according to its conformity with the existing value system.

7.   If step six is carried through successfully, the acceptance and gradual routinization of the new system into new patterns of performance and action. The new becomes the normal.

Figure 1 provides a model of the action stages described in this section.

The questionnaire used with the systems designers and managers followed this model covering a discussion of values, goal setting and goal attainment with particular emphasis on the systems design process; mechanisms, including implementation strategies for ensuring adaptation to the new system; an evaluation of the successful integration of the system in operation, and a description of pattern maintenance processes directed at its continuing effective operation. Questions were generally open-ended and where scales were used in the value questions the systems designers and managers were asked to give reasons for choosing particular points on the scale.

Figure 1. Framework for analysis of case studies

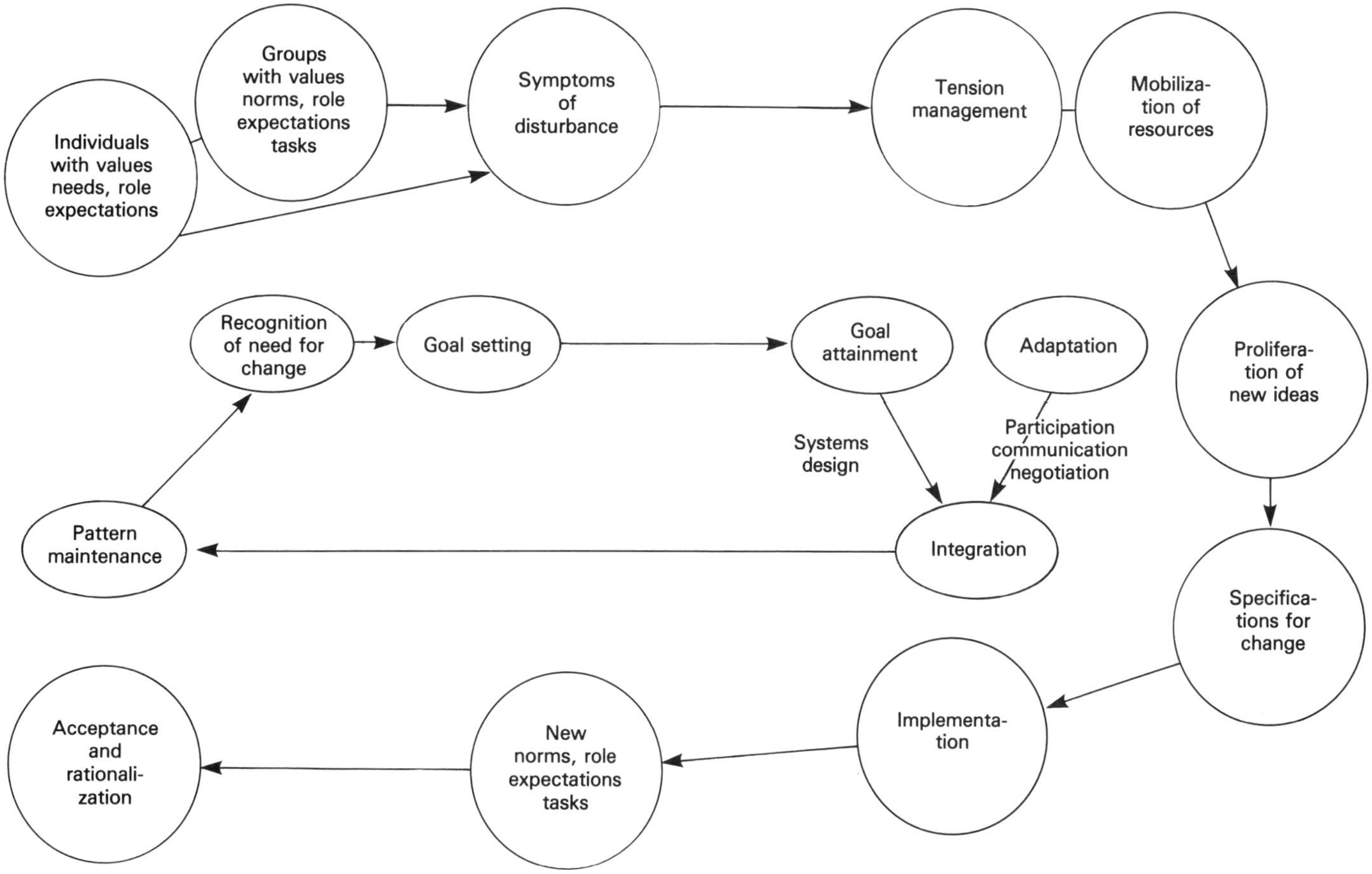

6. The Two Industrial Firms:
A. Chemco Distribution Services Department

In the two industrial firms the researcher had the advantage of being able to make before and after comparisons in the same departments. In each of these situations the impact of a batch computer system on job satisfaction and job content was observed and then, two years later, the consequences of a change to an on-line system were recorded.

The distribution department in Chemco is split into two sections: home trade and export trade, and controls the distribution function at five works producing a wide range of products. Certain products are produced by one works (single-works products), and others by more than one works (multi-works products). Each works has a distribution officer functionally responsible to the central distribution manager for the distribution activities within that works, with a staff which varies in numbers according to the size of the works.

All orders are routed through the distribution department located in Chemco. Despatch documents are raised, checked and distributed to the appropriate supplying works.

In 1975, when the first stage of this research was carried out, the home section of the department had recently changed from a batch to an on-line computer system. The export section had a batch system but was proposing to transfer to an on-line system in the near future. The discussion here covers the design of the home trade section on-line system and a post-change evaluation of its efficiency and job satisfaction consequences. Also, a description of the design of the export section batch system, an evaluation of its consequences and a follow-up study to examine the impact of the on-line system on clerks in the section. This second stage of the research was carried out in 1978 when the on-line system had become operational.

6.1. HISTORY OF THE HOME TRADE SECTION BATCH COMPUTER SYSTEM

The home trade section of the distribution department handles a variety of products including soda ash, sodium carbonate, caustic soda and salt. Prior to the introduction of the batch computer system in the late 1960s orders were processed manually using a master document which carried all the constant information about customers, from which despatch and invoice documents could be reproduced with the minimum additional entries. The basis of this system was a translucent order card (TOC) which could be easily photocopied. These translucent order cards were produced for each unique combination of customer/product/package/size of lot/price. On receipt of a customer's order at a sales office the clerk extracted the appropriate translucent card from his file and entered details of the order upon it. A photocopy was then sent to the distribution department at Chemco if the order was for a multi-works product, and from there it was routed to the works. In the distribution department a clerk calculated the invoice value of the order and had the customer's and internal invoice copies produced. When the goods had left the works for the customer, the works distribution clerk sent a photocopied order/invoice master to the invoice section in the Chemco home sales section.

The batch computer system effectively transferred the master (TOC) document to the computer file where it was held on magnetic tape. In the computer it again contained one record for each combination of customer/product/package, etc., each record being identified by a unique master file number. The computer also held a name and address file, a product/package file containing full descriptions of products and packages and an "outstanding order" file which contained the records of those orders which were in the process of being completed.

Translucent order cards (TOCs) which were printed records of the information stored in the computer were held by the sales distribution office responsible for receiving and processing customers' orders and were used to raise orders prior to their processing by computer. Once an order clerk in a sales office had completed a TOC it was passed to a teleprinter operator for transmission to the computer. Orders raised in the Chemco distribution department were multi-product or special orders and these were coded and punched in the normal way. The computer produced despatch sets; these were split in the distribution department and a copy sent to the works. Once orders were despatched details from "advice to invoicing point" (AIP) forms were entered into the computer. The computer then prepared and printed the invoices and entered details on its file. The computer also checked that an invoice had been printed for every original order which it had received and entered on the "outstanding order file" at the processing stage.

In 1972 investigation into the conversion of the batch system to on-line processing began. A feasibility report was issued in January 1973, followed by a project report in September 1973 which recommended that:

> The computer system be converted to on-line processing, supporting terminal equipment at the various Distribution Departments in the Company, and accepting continuous order entry from Sales Office teleprinters.

The system would require the installation of visual display units (VDUs) in the user departments for the input of data to the computer, and terminal printers for the continuous production of despatch documents. This would enable users to be in direct and continuous communication with the computer, eliminating the time-consuming and expensive process of sending all documents to the computer centre for conversion to computer input.

The system would process all home trade orders, including the printing of despatch sets and invoices arising from these orders, for all despatches of finished products from all works and depots of Chemco. The four files – TOC master, customer name and address, product/package and outstanding orders – would continue to be used in the on-line system. The batch system was converted to on-line operations in 1974 with a phased introduction of VDUs.

6.2. THE HISTORY OF THE EXPORT SECTION BATCH COMPUTER SYSTEM

The handling of export orders by the export section of the distribution department covers the receipt of orders, order processing, arranging shipping, transportation of goods to the docks and the preparation of invoices. The section has to communicate with the works and with customers and its business is mainly with overseas subsidiaries of Chemco. The work is complex and requires a great deal of documentation.

In 1971 a feasibility study was carried out to see if the company's computer facilities could be used to assist the work of the export section. This study suggested that a computer could be used to advantage in the two primary activities of the section, the conversion of customers' orders into the required documentation with all the data required for despatch, and the preparation of invoices after despatch. Such a system would provide both financial savings and easily accessible product market statistics. Following the model of the home section computer system, it was suggested that the computer should hold the master records and print out copies of these for order clerks so that they would have a visual record of the contents of the computer file.

Because of the complicated requirements of export invoicing it was proposed to first computerize the ordering procedures and to delay computerization of invoice procedures until a later date. It was hoped that eventually the computer would calculate and print the invoices and credits from the despatch information. This was an extremely complicated area and it might only be possible to do this for the major markets. The system would cater for all normal despatches of products from works or depots, at home and overseas, although urgent orders would still have to be processed manually. The first stage of this system was implemented in 1972 and the second stage in 1973. This system was converted to an on-line system in 1976.

6.3. ORGANIZATIONAL VALUES

The systems designers, the senior management of the distribution department, the home and export managers and the clerks in the two sections completed the "values" questionnaire in 1975 when the home sales section was using an on-line and the export section a batch computer system.

All three groups saw the values of the firm as in the centre or to the left of the scale, with an emphasis on discipline, standardization and efficiency. The only

Figure 1. Perceived organizational values (modes)

Place a tick where you think this firm fits on the scale below.

It has a strong belief in the importance of shared values – that employees should agree with its objectives and the way it carries out its operations

It likes to have disciplined employees who will put the firm's/office's interests first and willingly accept orders and instructions

It prefers to use standardized methods and procedures whenever possible as it believes that this assists efficiency in this kind of firm/office

It places a great deal of emphasis on efficiency and high production and less on personal qualities such as friendliness, trustworthiness, cooperation etc.

It feels the need to organize work activities into tightly structured jobs which are clearly defined and do not permit a great deal of individual discretion

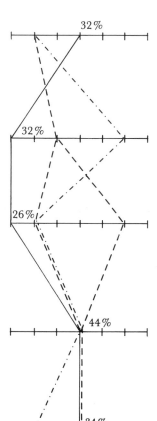

A multiplicity of different values exists and is tolerated here. It is quite acceptable for employees to hold very different views on what the firm should be doing and how it should be doing it

It does not believe in very strict discipline and tries to provide a situation in which people can pursue their own interests

It likes employees to work out their own methods for doing things and to use their own judgement when taking decisions

It places a great deal of emphasis on personal qualities such as friendliness, trustworthiness, cooperation etc. and less on efficiency and high production

It tries to organize work in such a way that employees have loosely defined and structured jobs which permit a great deal of individual discretion

——————— N = 22   Clerks
– – – – – N = 2   Managers
· – · – · – N = 2   Systems designers

Figure 2. Desired organizational values (modes)

Please tick where you *would like* this firm to fit on the scale below.

<table>
<tr>
<td>

It should have a strong belief in the importance of shared values – that employees should agree with its objectives and the way it carries out its operations

</td>
<td>

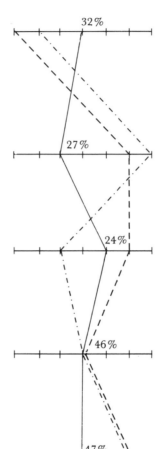

</td>
<td>

A multiplicity of different values should exist and be tolerated here. It should be quite acceptable for employees to hold very different views on what the firm should be doing and how it should be doing it

</td>
</tr>
<tr>
<td>

It should try to create disciplined employees who will put the firm's/office's interests first and willingly accept orders and instructions

</td>
<td></td>
<td>

It should have an easy-going kind of discipline and try to provide a situation in which people can pursue their own interests

</td>
</tr>
<tr>
<td>

It should try to use standardized methods and procedures whenever possible as this would assist efficiency in this kind of firm/office

</td>
<td></td>
<td>

It should encourage employees to work out their own methods for doing things and use their own judgement when taking decisions

</td>
</tr>
<tr>
<td>

It should place a great deal of emphasis on efficiency and high production and less on personal qualities such as friendliness, trustworthiness, cooperation etc.

</td>
<td></td>
<td>

It should place a great deal of emphasis on personal qualities such as friendliness, trustworthiness, cooperation etc. and less on efficiency and high production

</td>
</tr>
<tr>
<td>

It should try to organize work activities into tightly structured jobs which are clearly defined and do not permit a great deal of individual discretion

</td>
<td></td>
<td>

It should try to organize in such a way that employees have loosely defined and structured jobs which permit a great deal of individual discretion

</td>
</tr>
</table>

——————— N = 22   Clerks
– – – – – – N = 2   Managers
·–·–·–·–· N = 2   Systems designers

102

Figure 3. Organizational models (modes)

Please tick on the scale below what you believe to be the best form of department structure for clerical staff in general.

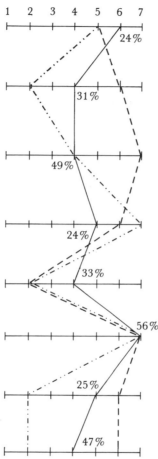

Jobs should be clearly defined, structured and stable	Jobs should be flexible and permit group problem solving
There should be a clear hierarchy of authority with the man at the top carrying ultimate responsibility for all aspects of work	There should be a delegation of authority and responsibility to those doing the job regardless of formal title and status
The most important motivators should be financial, e.g., high earnings and cash bonuses	The most important motivators should be non-financial, e.g., work challenge, opportunity for team work
Jobs should be carefully defined by O&M department, management services or supervision and adhered to	The development of job methods should be left to the group and individual doing the job
Targets should be set by supervision and monitored by supervision	Targets should be left to the employee groups to set and monitor
Groups and individuals should be given the specific information they need to do the job but no more	Everyone should have access to *all* information which they regard as relevant to their work
Decisions on what is to be done and how it is to be done should be left entirely to management	Decisions should be arrived at through group discussions involving all employees
There should be close supervision, tight controls and well maintained discipline	There should be loose supervision, few controls and a reliance on employee self-discipline

———— N = 96   Clerks
— — — — N = 3   Managers
— ·· — ·· — N = 3   Systems designers

103

Figure 4. Models of man (modes)

Please tick on the scale below how you see the general characteristics of the majority of staff in this department.

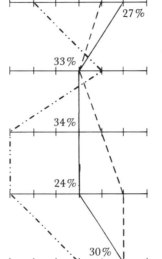

They Work best on simple, routine work that makes few demands of them	They Respond well to varied, challenging work requiring knowledge and skill
They are Not too concerned about having social contact at work	Regard opportunities for social contact at work as important
They Work best if time and quality targets are set for them by supervision	Are able to set their own time and quality targets
Work best if their output and quality standards are clearly monitored by supervision	Could be given complete control over outputs and quality standards
Like to be told what to do next and how to do it	Can organize the sequence of their work and choose the best methods themselves
Do not want to use a great deal of initiative or take decisions	Like, and are competent to use initiative and take decisions
Work best on jobs with a short task cycle	Able to carry out complex jobs which have a long time span between start and finish

—————— N = 96 Clerks
· — — — — N = 2 Managers
— ·· — ·· — N = 3 Systems designers

104

exceptions were the systems designers, who did not think the firm believed in strict discipline and a disagreement amongst the managers on whether the firm liked standardized methods or preferred employees to work out their own.[1] All three groups tended to be in the centre or to the right of the scale on what they would like the values of the firm to be, but with the systems designers favouring more standardization of methods or procedures than the managers or clerks. These answers suggest that the systems designers and the distribution department management and staff would like the firm to move to a more theory Y-value position than they perceived it as holding in 1975. The home and export managers suggested that the firm was already moving to a less tightly structured form of organization and that, although there were closely defined job responsibilities, within certain limits an individual was free to perform his job in the way that he considered best. Providing work quotas were completed, clerks in the distribution department had a great deal of discretion in how they programmed their working day, and the introduction of flexitime had added to this freedom.

Clerks were consistently in the centre or on the right-hand side of the scale in answer to the question on "the best form of department structure for clerical staff in general", liking flexibility and opportunities for self-management. The management value position was even further to the right than that of the clerks, although management believed that targets should be set and monitored by supervision. The systems designers held a more theory X-viewpoint favouring a clear hierarchy of authority, decisions by management and close supervision.

The managers thought that there was considerable untapped potential amongst their staff. They suggested that closely defined and oversupervised jobs tended to stifle initiative, reduce cooperation and remove satisfaction from the better employees. One of them favoured loose job definitions which could provide job satisfaction and financial rewards for those who wished to be high achievers. They also described the advantages of unstructured work groups in which the members had the responsibility for distributing tasks to those with skills and knowledge to handle them. The export manager hoped that the change to an on-line computer system would provide an opportunity for removing some of the existing job interfaces and creating an environment in which individuals could carry out a wider range of duties, with rewards more closely linked to achievement. The managers and clerks saw the characteristics of the clerks as fitting this value position, but the systems designers were on the left-hand side on four of the seven variables. They saw the clerks as having more limited talents.

The value scales suggest that clerks and their managers are in close agreement on many aspects of the work environment, with the managers favouring an even more flexible work situation than the clerks. The systems designers do not see

---

1. The mode is difficult to identify where there are very small groups. If all the respondents gave different answers then the answer that is nearest the ends of the scale is marked. This is justified on the grounds that the research is more concerned with extremes and differences than with neutral positions and similarities.

things in the same way and regard an undemanding, structured environment with tight controls as more appropriate.

6.4. WORK VALUES

The systems designers and managers were asked what factors they saw as most important in producing efficiency, happiness and a sense of fairness and justice in an organization. Their answers are shown in Figure 5.

Efficiency is seen here as a consequence of rational management policies, although there are suggestions that it is assisted by a number of emotional (cathectic) responses on the part of employees, particularly those associated with commitment and feelings of contribution.

Happiness is also seen as related to feelings of contributing and of being rewarded for good performance, and it is suggested that to stimulate these the organization must provide cognitive (intellectual), cathectic (emotional) and moral inducements. Fairness and justice is described in terms of equitable personnel policies and structures but also requires certain attitudes and responses from management. It is interesting that in answer to all three questions the cathectic responses – those related to attitudes and feelings associated with pleasure or attraction – come mainly from the department managers, who seem to have a strong identity with the psychological needs of their staff.

Efficiency, happiness and a sense of fairness and justice all contribute to employees' job satisfaction and so the systems designers and managers were asked how they would define this term. The systems designers' definitions were:

—Not being disturbed, doing what you are used to doing.
—Receiving a paycheque at the end of the month and thinking ''that's a lot of money but I earned it''.
—Also, not being ashamed of what you do.
—Recognition of the value of one's job (a) by the individual, (b) by the management.

The managers' definitions were:

—To have done something better than anyone else could have done it.
—Job satisfaction will result from a person feeling that the work he does regularly tests his initiative and ability, that he performs this work competently and that his competence is recognized by the management in practical and financial ways.

Because this research is concerned with relating personal values to the processes of systems design, both managers and systems designers were asked to list and rank in order of importance the set of activities which they included in the phrase ''system design''.

Figure 5. Chemco. Work values (systems designers' and managers' view)

EFFICIENCY

Role expectations
(what is expected of staff)

Cognitive	Cathectic	Moral
	A belief that they are doing a worthwhile job. A recognition that they are making a worthwhile contribution. A commitment to the objectives of the organization.	

Role requirements
(what the organization must provide)

Cognitive	Cathectic	Moral
Good management* Good management controls* Good management decisions* Good communication lines* Jobs which are necessary and useful* Well-designed systems* Organization flexibility Salary structures to attract the most able Career planning to identify and promote the most able	Sufficient internal friction to avoid complacency.	Sensitive and consistent personnel policies.

HAPPINESS

Role expectations
(what is expected of staff)

Cognitive	Cathectic	Moral
	A feeling of contributing to the total business. A feeling that exceptional performance will bring high reward.	

Role requirements
(what the organization must provide)

Cognitive	Cathectic	Moral
An adequate volume of work* Challenging work* Money*	A recognition of the individual and his worth.	An equitable distribution of work*

(Figure 5 continued on page 108)

(Figure 5, continued from page 107)

FAIRNESS AND JUSTICE

Role expectations
(what is expected of staff)

Cognitive	Cathectic	Moral
	Job satisfaction*	

Role requirements
(what the organization must provide)

Cognitive	Cathectic	Moral
	Responsive management*	Fair promotion oppor-
	Sympathy for personal cir-	tunities*
	cumstances	Equitable distribution of
	Awareness of the individual	wealth (profit sharing)*
	Willingness to provide job	Fair grievance procedures*
	interest	Acceptable communication
		and consultation procedure

* indicates the suggestion of a systems designer

The home and export managers provided the following lists of what they considered to be system design activities.

Manager A	Manager B
Understanding the business area	Ensuring operating advantages from the
Appreciating the aims and objectives	system
of the proposed system	Gaining job enrichment
A commitment from the responsible	Ensuring cost/benefits
manager	Anticipating future trends in the busi-
A passing on of that commitment	ness
to the staff	Ensuring the compatibility of the sys-
Employing a competent and knowledge-	tem with other internal/external sys-
able team	tems
Testing all proposals	Ensuring that it permits flexibility in
	the use of staff
	Ensuring that it caters for the needs
	of other departments as well as the
	needs of the ''user''

The lists of the systems designers were:

Systems designer A	Systems designer B
Output design	Understanding the user
Input design	Identifying the user's problem
Report writing	Knowing the objectives of the company
Presentation and discussion of report	Having a philosophy, a reason for
Discussions with senior programmer	doing things
Ideas formulation	Self-confidence

108

Specification writing   An ability to recognize nonvalid infor-
File design   mation
              Technical expertise

Systems designer C
Defining the user requirements (output)
Specifying user involvement (input)
Relating the system to user capabilities (staff qualities)
Defining the user training needs
Developing an implementation schedule
Relating this to system capabilities (hardware/software)
Evaluating additional hardware/software
Designing the computer system

It has been said that there is as yet no theory of systems design [Bjorn Andersen 1977; Mason 1973; Mumford 1966] and that today's approach is based on empirical experience of what seems to work. The different points of view which appear in these lists provide support for this belief.

The home and export managers described facets of systems design which would assist the integration of a computer system with the requirements of the user department and its surrounding organizational environment. The systems designers focussed more on their ability to understand and define the user's problems and on a set of sequential steps and procedures that would assist the development of an operational system. The impression given by the latter is of a somewhat ''us and them'' approach rather than a belief that the technologist and user should work together as a team jointly defining and solving problems.

6.5. GOAL SETTING AND ATTAINMENT

6.5.1. The Design of the Home Sales Section On-line Computer System

The home sales section in the distribution department receives and processes home trade orders. Order clerks are responsible for booking transport as well as processing orders and this is an interesting and complex part of their responsibilities. When an order is received it is put into the computer which then produces a standard despatch set appropriate to the product that has been ordered. This tells the works and the haulier that this order is required by this customer on this date. The order clerk fits the information on this despatch set into a plan which he prepares on a daily and weekly basis. He uses this plan to book the required transport and then sends it to the works. The works accept the plan by sending back a completed copy of the planning sheet together with one copy of the order known as an AIP (advice to invoicing point). This AIP is put into the computer and an invoice is produced for the customer.

The on-line system enhanced an existing batch computer system and was designed to process customer orders received in eleven sales offices and also those received directly by the distribution department home sales section. These last came

Figure 6. Chemco. Home order section workflow, on-line computer system

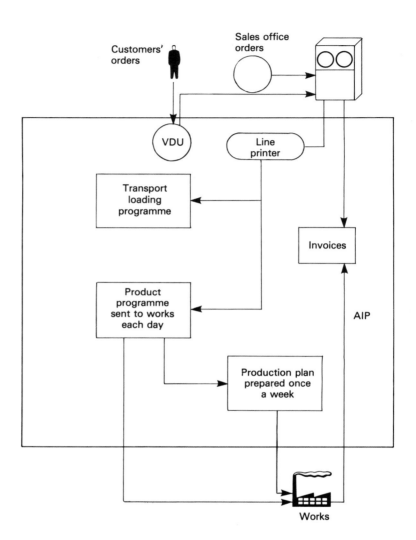

mainly from major Chemco customers and from other Chemco departments. Orders were inputted via VDUs and were printed out continuously, thus avoiding the time lag that had occurred with the batch system. When the feasibility study began in 1972, its terms of reference were, first, to investigate if an on-line system would process orders more quickly than the manual system and second, to establish if there would be cash flow advantages from such a change. Other objectives were to improve the planning of packaging for goods and the use of transport vehicles. A similar system was already working successfully in another division of the company and this provided a model for home sales.

The technical objective of the system was to obtain experience with on-line systems in an area where the user was interested in this kind of development and where financial savings were likely. The computer manufacturer, IBM, also had an interest in helping to develop an effective on-line system. The principle objective was to improve the company's cash flow position by sending out invoices to customers more rapidly. No specific human objectives were set at this stage. VDUs were new and their human consequences still relatively unknown, although it was thought that the fast feed-back on errors which they could provide would lead to greater work accuracy. These objectives were jointly agreed by the systems designers and the user management. The costs of the systems were covered by the reduction in staff numbers that the batch system had already secured.

Interest in a change to an on-line system came initially from the user managers and they participated in the design process from the start of the project. A pre-feasibility report on the advantages of on-line operation was prepared by the distribution manager and the feasibility team consisted of a representative from management services, the distribution manager and the sales organization manager. This team had also prepared the feasibility report for the batch system. The feasibility study for the on-line system took eight months and was completed in 1973. As part of its work the team was given an additional brief, that of examining whether sales offices were essential or whether orders could be routed direct from customers to the company so that there was only one order reception point. After investigation it was decided to retain the sales offices as they did a great deal of business with small firms such as dry-cleaners, but a result of this new brief was that the feasibility study incorporated a discussion of organizational issues as well as consideration of a new level of computerization. However, because the batch system was already in existence, fundamental changes in the sales concept could not be considered and it was decided to modify and enhance the existing system. Once the feasibility study was accepted a cost-benefit analysis was made which showed a nominal saving of about £2000 a year.

The design of the system was not subject to any major constraints. IBM already supplied computers to Chemco and it was logical to stay with one manufacturer and purchase IBM terminals. These terminals were expensive and this fact led to a modification of the original idea of inputting both orders and AIPs (advice to invoicing point) via the VDUs. It was decided to input order information only, in this way reducing the number of VDUs that would be required. Financial constraints also led to a decision not to install VDUs in the sales offices and AIPs continued to be sent by post.

6.5.1.1. *Design Alternatives*

Because the on-line system was an enhancement of the existing batch system, no technical alternatives were considered, and technical development consisted of converting the batch system to on-line with VDUs for order input. The technical problem was primarily one of how much information to transmit on-line and as this was a pilot system it was thought safer to initially restrict this to essential information and leave the rest, for example invoice information, until later.

Human design alternatives were concerned with who should operate the VDUs. One approach was to have order clerks and VDU operators working in teams of two, with job rotation; a second was to have trained VDU operators in a specialist section; a third to let all staff in the office operate the VDUs. The advantage of the first alternative was that it avoided physical problems such as eye strain that might occur with constant machine operation, but it introduced a new position into the existing work organization. The second alternative had the advantage of speed and efficiency of trained operators, but fast order input was not critical to the efficiency of the department. It had the disadvantage that two groups would be involved with the input of information, order clerks and VDU operators and this might cause problems. The solution chosen was to site the VDUs in the home sales office and allow everyone there the facility of using them. The philosophy behind the choice was that the order clerks had responsibility for responding quickly to orders, therefore they should be able to make this response themselves. But anyone who did not wish to use the terminals had the option of not doing so.

This decision was arrived at through discussion between the systems designers and the home sales section management, with the user making the choice. The volume of work, around fifty orders a day per order clerk, made such a solution quite feasible.

The home sales computer system is an example of a good batch system replacing a good manual system and a good on-line system replacing a good batch system. The care that had always been taken in designing effective work systems enabled each level of technology to be enhanced without difficulty.

6.5.2. The Design of the Export Sales Section Batch and On-line Systems

The design of these two systems was considerably influenced by the nature of the distribution activities of Chemco. The products are mass-produced bulk chemicals such as sulphuric and hydrochloric acid which are difficult to store and therefore have to be moved to the customer as quickly as possible once they are produced. This means that there is little production planning and products are made in response to customer orders. Despite this dependency on the market, production is fairly stable with the same customers placing regular orders. A great deal

112

of business comes from other manufacturing industries who telephone in and say how much of a particular product they require. The work of distribution is therefore primarily concerned with order processing and invoicing and not with production planning.

Export clerks receive orders from customers who are mainly Chemco subsidiaries in Europe and elsewhere. They check the order, validate prices, check licences for import, establish that the customer can pay and then produce a set of order documents. These are sent to the works with a shipping date, to the shipping office in Liverpool who will book space on a ship, to the works transport department and to the customer giving the expected date of delivery. In addition clerks prepare invoice documents. They therefore have complete responsibility for the processing of an export order without the intervention of a third party such as a marketing department. This means that their work must be extremely accurate, otherwise the company could lose money.

With the manual system export orders were handwritten and then typed, a method that was slow and expensive. It was thought that the introduction of a computer would reduce staff requirements, ensure a smoother processing of orders and provide a bank of knowledge which could be held in the computer. When the first feasibility investigation was commenced in 1971 the home trade section already had a batch computer system that was working successfully and providing a great deal of useful statistical information, and this provided a stimulus to go ahead.

The technical objective for the batch computer system was primarily to provide a system similar to that used by home trade. Business objectives were to process orders more efficiently, obtain more accurate information and eventually to carry out production planning on the computer. No human objectives were specified but there was a desire to use the computer to remove routine work from the clerks. It was proposed that the system should have two design stages: first order processing and later invoice processing. In fact invoice processing was never carried out by the batch system as it was recognized that an on-line system was a logical prior step. Because of the complexities of export orders it was difficult to change the existing organization of work in any major way and the computer mirrored the manual system with computer-printed instead of typed documents. The design of forms proved particularly difficult as it was not easy to create standardized documents for such varied work.

The user manager was enthusiastic about the proposed system, especially the predicted reduction in costs of £13,000 per annum. In fact these were not achieved as most of them stemmed from the computerization of invoicing which was delayed until the introduction of the on-line system, but the system did break even financially.

The export clerks were sceptical about computerization. Although they did not object, there was a feeling that their work was too difficult and complex to be handled by a machine. This reaction influenced the management services department to produce a simple system that did not incorporate any revolutionary new thinking. Despite this aim, the export clerks saw the batch system as imposing

113

Figure 7. Chemco. Export sales section workflow, batch computer system

Department staff are one manager, two assistant managers, four group leaders, order clerks, invoice clerks, document disposal clerks, general duties clerks, order writers, typists.

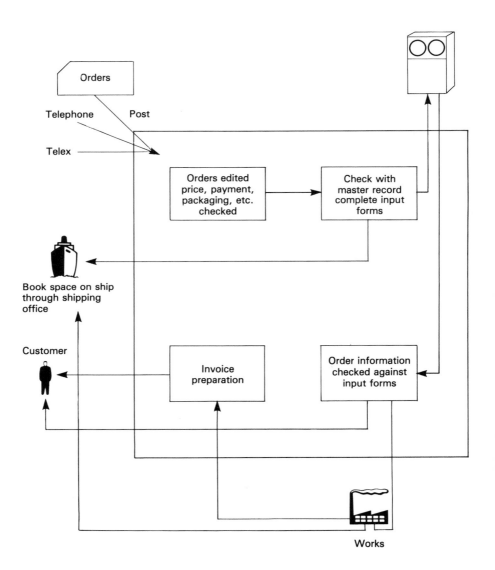

certain constraints on them. For example, many orders received from overseas were not valid until the letter of credit arrived and therefore could not be put into the computer immediately. This meant that information on outstanding orders derived from the computer was not completely accurate. Another constraint was that the volume of information was more than the computer could cope with and orders were not processed until three months or less before the delivery date. This meant that orders for twelve months of deliveries had to be reinputted every three months.

Four factors encouraged a change to an on-line system. First, the export sales clerks were not happy with the system and complained about it to management services; second, marketing and shipping departments were not getting accurate information from the system because orders were not always put immediately into the computer in case amendments came later; third, the manager of the export sales section was anxious to have an on-line system; at that time (1973) the on-line system was being designed for the home sales section.

It was hoped that an on-line system would do the following:

1.   Enable order information to be put into the computer as soon as it arrived in the division.
2.   Keep information as accurate and up to date as was feasible.
3.   Provide the user with better documentation.
4.   Provide the up-to-date information necessary for invoices.

The systems designers were also hoping to reduce the complaints of the clerks about the computer by providing them with a more flexible system.

Like the batch system the on-line system was not easy to design. For example, a very large amount of documentation was required and single orders could contain a considerable amount of data. Also, while the life of an order was about six weeks, it could be amended a number of times between receipt and despatch. The system was expected to provide some financial benefits and a target of £14,000 saving per annum was set.

6.5.2.1. *Design Alternatives*

Few technical alternatives were considered when the export batch system was designed for both the user and management services agreed that the best strategy was to computerize the existing manual system. There was some discussion on whether to print out a master set of documents and then amend these to fit different sets of needs or to print out a number of different documents each providing different kinds of information. It was eventually decided to do the latter and to print four different sets of documents for each order. This solution had the advantage of being preferred by the export clerks but the disadvantage of requiring the computer to be loaded with a number of different forms.

A new development that might have been considered at the time was the automatic updating of prices. This would have helped the export department in the calculation of prices and avoided the necessity for holding prices on the master record. It had the disadvantage of requiring organizational changes, for the export order clerks were located in the distribution department, whereas the export department was at another Chemco location in ten miles away. Automatic pricing would logically require the order clerks to join the export department. The principal factors influencing the choice of technical solution for the export section batch system were the level of technology available at the time, and the ease of getting the system accepted by the user. To be acceptable it needed to be non-controversial, simple and easy to introduce.

When the system was enhanced to on-line, the principal technical consideration that had to be investigated was: should the existing batch system be completed by adding to it the processing of invoices, before moving to an on-line system? Or should the move to an on-line system precede the processing of invoices? The advantage of the first option was that this would be a smaller problem for management services to tackle and also less expensive; the disadvantage was that the user preferred to move directly to an on-line solution. It was decided to move to an on-line system immediately. Another technical problem concerned documentation. With the batch system 17 different items of paper were produced when the order entered the system and there was a great deal of discussion about how this documentation could best be organized.

A technical option that might have been considered was the elimination of the master record and the replacement of this with a customer file and a product file which would be brought together when an order was processed. The advantage of this was that there would be standard information for each customer, the disadvantage that records changed frequently in the export section and the elimination of the master record would cause the export clerk a great deal of additional work. The principal factor influencing the choice of technical solution for the on-line system was a desire to make the input and amendment of orders easy for the user.

Human alternatives that were considered with the batch system concerned the organization of the export sales section. Management services suggested that the section should be split into an invoicing group and an order group, their argument being that the reduction in staff brought about by the computer would not warrant an invoice clerk in each market group. At the time the section was organized into four market groups: (a) home trade and Europe, (b) the Americas, (c) the Middle East and Africa, and (d) the Far East. This proposal was not acceptable to the export section.

With the on-line system the question was whether to have specialist VDU operators or give the order and invoice clerks responsibility for handling their own VDU input. It was decided not to have specialist operators. The present situation is that in some market groups the order clerks input their own orders, while in others they are assisted by input clerks who combine order input with some general clerical duties.

Because the philosophy of the batch system was to computerize the manual system, not a great deal of attention was paid to the design of jobs. The order clerk found that he had a new element of routine introduced into his work as he now had to complete a detailed form before the order was coded and punched for

116

input to the computer. Some order clerks disliked this task and so the extremely routine role of order writer was introduced into some market groups.

The introduction of the on-line system removed this routine but a number of clerks thought that even using the VDUs for order input interfered with their main task of providing an efficient customer service. The view of the systems designers was that their system ought to be flexible and permit a choice of work organization and that the user should be helped to make his own choice. In order to assist this flexibility all members of the export sales section were given or offered training in the use of VDUs.

6.6. ADAPTATION

Adaptation involves providing the user with a commitment to the new computer system, with an understanding of how it works and with the knowledge and skills to operate it efficiently. Commitment is facilitated by involvement in the design processes and by a belief that the system will fit user needs and provide an improvement on the previous work method.

Adaptation in the home sales section was assisted by the regular progress meetings of a committee which met once a month to discuss the computer system. This committee had a flexible membership with both the systems designer and the home sales manager bringing people to meetings who had an interest in the subject being discussed. Its role was to review the computer situation and decide priorities and in doing this it made a usful contribution. But the clerks as a group had a limited formal involvement in the design of the system, restricted to discussions about their work, although the systems designers thought that they succeeded in developing many close informal contacts with them.

With the export batch system no formal consultative structure was created but there were regular meetings between management services, the user manager and group leaders in the section. More consultation took place with the design of the on-line system and there were weekly meetings over a period of two months, each attended by section leaders and two of their order clerks. At these meetings different aspects of the system were discussed by the systems designer, and attention was given to methods of implementation. The systems designers thought that these meetings had a positive effect on the morale of the export clerks and made them feel involved with the new system. Unfortunately, they ran out of steam and did not continue for very long. Adaptation was therefore excellent at management level with a great deal of involvement of the home and export managers in the design of the systems. It was less effective at the clerical staff level, relying on consultation and communication rather than any real participation in systems design. This had no disadvantageous consequences in home trade but may have had in export where initial difficulties required a high level of commitment and tolerance from staff.

6.7. INTEGRATION

Successful integration requires a good fit in terms of efficiency and job satisfaction between the human and technical parts of a system with people believing that the computer assists their efficiency and increases their job satisfaction, the two being related. It also requires a good fit between the internal and external environments with the new man-machine relationship contributing to a better external relationship such as an improved customer service. Research suggests that job satisfaction is influenced by the task structure of work and so before discussing the level of job satisfaction in each section the content of the different jobs will be briefly described [Herzberg 1966; Cooper 1973; Taylor 1975].

6.7.1. The Distribution Department

6.7.1.1. *Structure of Work*

There were four principal jobs in the home sales section – planner/allocator, planner, general duties clerk and transport clerk. The planner/allocator had responsibility for maximizing the use of vehicles sub-contracted from hauliers for the transportation of different products. This involved the making of forward plans regarding the availability of products and informing customers of when they might expect delivery. The work of the planner was similar except that he prepared a loading programme for products but did not arrange transport. The work involved considerable interaction with other groups and many decisions, for example, on how to handle rush orders, cancelled orders, or situations when a customer could not accept delivery, for example because of storage problems. The general duties clerk assisted the planning and allocating functions, sending information to customers and checking output from the computer. This was a bitty and poorly integrated job. The transport clerk's task was to assist the transfer of orders from the works to the customer and to sanction road and rail movements made by the planner/allocator.

The jobs in the export section were export order clerk, invoice clerk, document disposal clerk, order writer and general duties clerk. The export order clerk had responsibility for processing orders and for ensuring that a high level of quality and service was provided to the customer. This involved dealing with customer queries, investigating why orders had not been despatched and liaising with the pricing section. The invoice clerk prepared and despatched invoices to customers and before the introduction of the on-line system had calculated the amount to be charged. Once the on-line system was introduced, these calculations were made automatically by the computer. The document disposal clerk had a document collecting, collation and dispersal function, sending documents to the appropriate recipients including customers, banks, and treasurer's department, etc. The order writer was an ancillary to the order clerk and only existed because some order clerks did not like the routine task of writing out orders for the batch computer system. The general duties clerk was responsible for a number of checking activi-

Table 1. Chemco. Job structure comparisons

CONTROL LEVELS

Definition of skills:
A. Communicating in writing
B. Communicating verbally
C. Arithmetical skills
D. Machine operating skills
E. Checking/monitoring/correcting
F. Problem solving
G. Coordinating
H. Supervising

These are the day to day operating tasks

L = Large number of tasks, M = Medium number, S = Small number.

Level	Group	Aspect	Sub-aspect
LEVEL 1 — OPERATIONAL LEVEL	Variety (Give sense of personal control)	Reduce boredom	No. of tasks / No. of skills / of methods / of work sequence
	Means choice (Give feeling of achievement)		Use of judgement
	Knowledge choice		Use of initiative / Clearly defined start to job / Clearly defined end to job / Uninterrupted task sequence / Long task cycle 20+ minutes
	Job identity (Give sense of making important contribution)		Visible contribution to product or service
		Give sense of team work	Works as member of group
	Job relationships	Give sense of confidence	Considerable inter-action required / Clear work objectives / Objectives not too easy or too difficult
LEVEL 2 — Goal clarity		Problem prevention	Can requisition required resources / Can correct errors, solve problems
LEVEL 2 — Anti-oscillation		Efficiency improvement	Coordinates own work activities
LEVEL 3 — Optimisation (HIGH CONTROL LEVELS)		Creativity	Coordinates group work activities / Can improve methods
LEVEL 4 — Development		Autonomy	Can improve product or service
LEVEL 5 — Overall control		Key task	Group free from supervisory control / Individual free from supervisory control

Job structure data

Job	Reduce boredom	No. of tasks	Skills	Key task
Export sales section. batch				
Export order clerk	L	6	A/B/E F/G/H	Order handling inc. transport. Problem solving
Invoice clerk	S	3	A/B/C	Invoice handling Checking
Document disposal clerk	S	4	A/B F/G	Document collection, collation, dispersal
Order writer	S	2	A/B	Prepares data input for computer
General duties clerk	S	1	A	Checking
Home sales section. on-line				
Planner allocator	L	4	A/B F/G	Order handling inc. transport planning
Planner	M	2	A/B	Planning
General duties clerk	S	2	A/B	Odd jobs
Transport clerk	S	1	A	Listing Calculating

Note: "Uninterrupted task sequence" for the Export order clerk is marked "oo"; "Uninterrupted task sequence" for the Planner allocator is marked "for some cases".

ties associated with the work of order and invoice clerks such as checking the calculation of invoices.

The author's impression at the time of her first visit to Chemco in 1975 was that both the home and export sales sections had too rigid and polarized a job hierarchy. The most responsible jobs of home sales planner/allocator and export order clerk were well-designed and met most of the criteria of a good task structure. Below this level, as Table 1 shows, the jobs became progressively more routine with, for example, general duties clerks having small jobs, with short task cycles and a poor level of task integration.

This organization of work was not greatly altered by the introduction of an on-line system into the export sales section, although, when the author made her second visit in 1978, the task structure of the export section was in the process of being reorganized so as to make it more flexible and satisfying for staff. The proposal was to organize each market section into teams, as shown in Figure 8.

This arrangement replaced a functional structure with order clerks, invoice clerks and document disposal clerks working in relative isolation from each other. The document disposal task and the invoice task, previously handled by two clerks, would now be combined and this new clerk would also operate the VDU. The order clerk would be relieved of using the VDU and this task would be carried out by an order input clerk. New staff would join the group as typists or order input clerks and would be able to move up the hierarchy as they gained the necessary skills and as vacancies arose. This meant that they would be experi-

Figure 8. Proposed organization Chemco export section

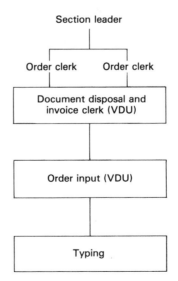

enced in many of the tasks associated with processing orders before they assumed the responsible role of order clerk. The combining of the document disposal clerk with the invoice task was seen by management as providing a larger and more interesting job, although the reaction of the clerks concerned was not enthusiastic. The advantage of this form of work organization is that it provides excellent training opportunities for each level of staff. Its disadvantages are that it is still hierarchical, with lower level clerks having to wait for vacancies in the grades above them, and it does not remove the very routine jobs at the bottom of the hierarchy.

Informal discussion with the clerks suggested that the order clerks found the on-line system an improvement on the batch system, but that the invoice clerks thought that the computer had removed some of the challenge from their job.

6.7.1.2. *The Clerks' Perception of Integration*

Tables 2 and 3 show the clerks' view of their job problems and their positive and negative feelings towards their work. Only the negative parts of the job satisfaction scales are shown in Table 2 as the aim is to identify work problems which might be related to the computer system. The data consists of the 1975 survey of home and export clerks' attitudes to their work and the 1978 survey of export clerk opinion. Comparisons can therefore be made between the 1975 attitudes of home sales clerks using an on-line system and those of export clerks using a batch system; and between the 1978 attitudes of export clerks using an on-line system and their attitudes in 1975 when using a batch system. A word of caution is required here. It cannot be assumed that attitudes are directly related to the computer system, for there may be other causes of content or discontent.

The impression Table 2 gives is of a higher level of job satisfaction in the home sales section, irrespective of whether export sales are using a batch or an on-line computer system. Fewer people in home sales saw work as usually or always routine; they perceived more opportunities for the use of discretion; more opportunities to see a piece of work through from start to finish, and felt freer of rules and procedures. They also appeared to be under less pressure with 58% (still a high figure) compared to over 70% saying that they had too little or barely enough time to do their work.

Despite the three year gap and the enhancement of the computer system, the two sets of figures for export sales are remarkably similar. Some staff saw the on-line system as having brought advantages, for 27% said their work now had more challenge, 17% said they could now take more decisions and 33% said that deadlines were easier to meet, but the majority view was that the computer had made no difference to work challenge or to the opportunity to take decisions and that deadlines were now more difficult to meet.

The measures of positive feelings to work also show the home sales section as more satisfied than the export group. A sense of achievement was felt more frequently, enjoyment of work was greater and 82% of staff saw their present job as approximating to their ideal job. In Table 3 the level of job satisfaction in export sales is again similar in 1975 and 1978.

Table 2. Chemco. Opinion on different aspects of work
(in percentages)

Category	Aspect	Statement	Type	Home on-line	Export batch	Export on-line
VARIETY	Routine and challenge	Work is usually or always routine	FACT	12	32	33
		There is too little challenge	PREF	20	21	20
		There is too much routine	FACT	30	56	53
		Almost everything has rules or set procedures	PREF	4	33	46
DISCRETION	Means choice	Can hardly ever choose our work methods	PREF	–	8	5
		Can hardly ever choose work methods	FACT	8	7	16
		Can hardly ever choose work sequence	FACT	9	20	18
		Can hardly ever choose next task	FACT	30	24	19
		Would like more choice of work methods	FACT	5	23	17
		Would like more choice of work sequence	PREF	9	23	14
		Would like more choice of next task	PREF	–	26	17
	Knowledge choice	Would like more opportunities to make decisions, use judgement	PREF	48	61	54
		Has very little or little chance to make decisions, use judgement	FACT	18	41	35
JOB IDENTITY	Task cycle and integration	Would like more chance to do this / see a piece of work through from start to finish	PREF	33	56	50
		Inefficient work by others can almost always or usually hinder me	FACT	71	72	69
		Inefficient work by me can almost always or usually hinder others	PREF	75	78	78
JOB RELATIONS	Work dependency	I would like there to be less work dependency	FACT	25	24	56
		There is too little or barely enough time to do work	PREF	58	87	74
GOALS	Deadlines		FACT			

Export section comparisons of on-line with batch system:

Category	Note	Value
VARIETY	But more challenge than before	27
DISCRETION	But more decisions now	17
JOB RELATIONS	Less	12
GOALS	But deadlines are easier to meet	33

122

Table 3. Chemco. Positive and negative feelings about work
(in percentages)

Negative feelings about work

	Time drags ¾ of the day or more (FACT)	My present job is far from my ideal job 1/2 (FACT)	I am very little involved in my job (FACT)	Taking the job as a whole I do not enjoy it (FACT)	I hardly ever get a sense of achievement (FACT)
Home on-line	4	–	–	–	–
Export batch	15	26	10	7	10
Export on-line	2	9	2	5	8
Export section comparisons of on-line with batch system	Time drags for me more often: 16			I enjoy it less now: 33	I get less sense of achievement now: 32

Positive feelings about work

	Time never seems to drag (FACT)	My present job is close to my ideal job 5/6 (FACT)	I am strongly involved in my job (FACT)	Taking job as a whole I enjoy it very much (FACT)	I would like to get sense of achievement more often (PREF)	I get sense of achievement almost every day (FACT)
Home on-line	65	82	33	58	58	79
Export batch	66	32	24	17	68	38
Export on-line	77	39	28	22	67	40
Export section comparisons of on-line with batch system	Time drags for me less often: 44			I enjoy it more now: 30		I get more sense of achievement now: 11

123

The clerks were asked whether the computer increased or decreased their efficiency and the interest in their work, and in the home sales section in 1975 answers were almost entirely in terms of an increase in efficiency and interest. Staff spoke of "easier access to information", "orders are received much quicker", and "it takes the donkey work out of the job". One clerk summarized the computer's advantages as,

> —It gives us more to do so that a person feels as if he/she is needed.
> —It makes us feel as if we have achieved something when the job is finished. It is really worthwhile.

Comments on the computer as an aid to efficiency in export sales were more evenly divided between favourable and unfavourable opinions. In the 1975 survey there were criticisms of the system as being too slow and needing too much preparatory paper work. In the 1978 survey there were rather more favourable comments. Order clerks saw the advantages of the system as:

> —It cuts down the time taken in doing the work. Also it spots mistakes quicker than the old system.
> —It gives you more involvement and interest in the work you are doing.

And its disadvantages as:

> —It is inflexible.
> —It inhibits by demanding too much detail which is not of benefit to the section.

More criticisms came from invoice and document disposal clerks:

> —The computer is geared to producing statistics and masses of paper and the requirements of the customer seem to take second place. Papers cannot be obtained as readily and any errors require more masses of paper to rectify.

The computer's impact on the interest of work was viewed more unfavourably in 1978 than in 1975 by invoice and document disposal clerks, although not by order clerks. This may have been a result of the computerization of invoices. Comments were:

> —The interest is much the same but the computer tends to make me cross and utterly frustrated because I never feel I am getting through the pile of paper on my desk.
> —It decreases interest because the computer does all the detail for you. It makes work very simple which is a help, but sometimes a drag.

From the clerks' point of view the integration of the computer into home sales appeared to have been very successful. It was less successful in the export section and the change from a batch to an on-line system did not seem to have improved the situation a great deal, although it assisted the work of the order clerk. Job satisfaction appeared higher in home sales than in export although it was not easy to identify why this should be so. Staff in export spoke of a need for more involvement, a reduction in work pressure and greater system flexibility.

6.7.2. Management's Perception of Integration

6.7.2.1. *The Fit with the Needs of the Clerks*

The systems designers saw the main strengths of staff in the distribution department as their long experience of the job and their ability to cope with routine work. Management saw them as a dedicated group, able to work cheerfully under pressure and loyal to their employer. Their weaknesses were seen by the systems designers as a tendency to concentrate on operational requirements, "getting the product to the customer" at the expense of computer requirements, "making sure the correct codes are used". Both systems designers and management saw the clerks as rather passive, wanting management to take their decisions for them and being reluctant to take advantage of opportunities for participation. This last criticism may not be a valid one, as in the author's experience white collar workers have to learn to participate and their ability to do so depends on the nature of the participative structures that are created.

There was some difference of view between systems designers and distribution management on whether the design of the computer systems had helped exploit these strengths and reduce the weaknesses. The systems designers thought that they had achieved both of these things and that the on-line systems helped correct weaknesses by bringing the computer closer to the user. Management thought that the opposite had been the case. The batch systems committed senior staff to routine work which could be carried out by junior staff and the introduction of prescribed routines and procedures tended to place staff in a straight jacket and reduce initiative. Nevertheless the introduction of more standardized procedures had improved work discipline. The management view was that when the batch computer systems were introduced an opportunity had been lost to make organizational changes which would have improved job satisfaction.

The home sales manager believed that the on-line system in his section had made a major contribution to the job satisfaction of his clerks, and his view was supported by the designer of the system who described it as

> —Much better than the batch system. Responsive to the clerks' needs and providing a reliable, prompt service. It is a good simple order processing system that eliminates form filling and enables orders to be easily amended.

The export sales manager was less certain that the batch system had improved the job satisfaction of his clerks but pointed out that it was difficult to differentiate between problems associated with the volume of work and problems associated with the computer system. He thought that some order clerks liked the system but that the clerks as a group did not, because of the large amount of document completion that was required and the fact that the system was no faster than the manual one. Adjustments had been made to the organization of work in an effort to ameliorate some of the disadvantages of the computer. For example, the job of order writer had been created to remove from the order clerk the chore of having to write out the orders before input to the computer. Markets which had similar kinds of documentation had been grouped together and when orders had to be amended the computer did not produce printouts until requested to do so. Both of these changes reduced the amount of paper arriving in the section.

The introduction of the on-line system had been made difficult through coinciding with the retirement of a number of experienced clerks in the section. New, young staff had been recruited as order clerks and they found the system difficult to operate with the result that there had been errors and problems. Partly in an effort to alleviate these difficulties a new role was introduced, that of input clerk. This new job had a similar responsibility to that of the order writer, except that orders were not put into the computer via the VDUs. A decision was also taken not to recruit directly into the order clerk position but to start people lower down the clerical hierarchy so that they gained a good understanding of the total system of work before being promoted to order clerks.

6.7.2.2. *The Computer's Contribution to Better Management Information and Control*

Management's view was that the computer systems had increased the amount of information available to the distribution department and this was supported by a comment from an export section leader in the 1978 survey who said

—The amount of information and statistics stored in the computer and readily available makes the job of supervising the performance of the group much easier.

For example, he was now able to obtain information on the location of demountable tanks, on sales into different markets and on outstanding orders for a particular commodity into a particular market.

Management could also obtain information on market demand and sales statistics and when production planning was transferred to the computer its information-providing potential would be even greater. The computer was also seen as increasing the timeliness of information and the on-line systems as producing a higher level of accuracy.

6.8. MANAGEMENT'S EVALUATION OF THE SYSTEM

The systems designers' evaluation of the two systems was that the home sales online system was successful because it had been designed with user requirements in mind and had succeeded in meeting these requirements; it had also removed the constraints of the batch system. The export sales system was seen as a good efficient system from the computing point of view, easy to maintain and easy for the user to understand. The distribution manager thought that the systems were successful because of the information they could provide on the movement of orders, and said they would be even more successful when linked into production planning. The home sales manager saw the speed of response of his on-line system as its major advantage and hoped that it would lead to the improved planning of production and vehicle use. The export sales manager welcomed the increase in information that he received, but said he had found the batch system too inflexible. For example, whenever an order was amended it produced a new set of documents which then had to be filed. This inhibited staff from putting in orders which might later be changed and this, in turn, caused the computer to produce inaccurate statistics. The on-line system was more flexible. Orders could be put into the computer via the VDUs even though some documents had not arrived, for example import licenses; whereas the batch system required all documentation to be complete. Amendments could now be made via the VDUs and printouts produced on request only. The on-line system was faster than the batch, and provided an immediate visual check of the accuracy of data input.

Both the systems designers and user manager agreed that the home sales on-line system had been highly successful in attaining the technical and business objectives that had been originally set for it. Although it was not possible to prove that the system had led to an improvement in cash flow, in the manager's opinion this must have happened. The export batch system was seen as fairly successful; it was more accurate and processed orders more rapidly than the manual system, but greater gains were expected from the on-line system. This was seen as having met the original objectives of a system that (*a*) would accept order information when it arrived in the company, (*b*) would keep information accurate and up-to-date, (*c*) would provide the user with better documentation, and (*d*) would provide the up-to-date information necessary for the production of invoices. However, the complexity of the data that had to be fed into the computer at the input stage still caused some problems. Also because orders were processed overnight in the computer, the paper did not arrive until the next day. This meant that very urgent orders still had to be processed manually.

The distribution manager's evaluation of the home and export sales systems was that they had led to better management control, a reduction in staff numbers, an improvement in cash flow, the better utilization of vehicles and greater job satisfaction. His only qualification was that instead of computerizing existing procedures there could have been more rethinking of organizational needs.

6.8.1. The Computer and Corporate Strategy

The distribution manager believed that the company still needed a more systematic method for evaluating the success of its computer projects, and for determining what computer developments should be embarked on next. But existing systems were greatly assisting the business strategies of the distribution department. There was now a faster picture and clearer definition of costs and the use of computers had improved the utilization of vehicles. In the future it was hoped to make the computer more of an aid to forward planning through sales and demand forecasting. This was already used in other parts of the company. The accounts department used the computer for forward financial planning and this included distribution. The computer was also used to assess the capital value of projects and to examine alternative strategies for the business.

In the past the success of computer policy had been expressed in terms of staff saving and since 1974 Chemco had quadrupled in size while staff numbers had been reduced from 24,000 to 16,000, which would not have been possible without the assistance of EDP. He hoped that in the future success would also be evaluated in terms of job satisfaction. He would regard a policy as having failed if it led to people being unhappy in their work.

6.9. CONCLUSIONS ON THE CHEMCO COMPUTER SYSTEMS

6.9.1. Goal Setting and Attainment

These were successful systems in their on-line form, for both technical and business goals were successfully achieved. The principal technical goal of the home sales system was for the systems designers to obtain experience in designing on-line systems, while the principal business objective was to improve the section's cash flow. The human objective, as seen by management services, was to provide the kind of system that met the user's needs and which the user could operate efficiently. All of these goals were achieved without difficulty.

The technical objective for the first export sales system was to develop an application similar to the batch system originally used by home trade, which had been liked by the clerks. Business objectives were to process orders more efficiently, and to obtain more accurate information. In addition, management hoped to remove some of the routine work from the section. Unfortunately the system was introduced at the time of an export boom when the staff were under considerable pressure of work. They found the input documents complex and time-consuming to fill in and disliked having to meet computer-set deadlines. One benefit of the system was that it transferred to the computer various routine calculations, but the clerks had not objected to doing these and they found the work of completing numerous input forms even more routine.

The on-line system largely remedied these defects by enabling order information to be put into the computer as it arrived in the section even though documentation was incomplete, it also reduced the amount of paper that the computer produced and increased the availability of information. But there continued to be input problems and some time delays which meant that very urgent orders had to be processed manually as with the batch system. This system had an unhappy start, for it was introduced at a time when the section was losing many of its experienced clerks through retirement and was bringing into senior positions clerks who had no experience of export sales. Once this difficult transition period was over the system settled down and largely achieved the objectives that were originally set for it. With hindsight management regretted that clearer human objectives associated with job satisfaction had not been set when both the home and export sales systems were designed and that the catalytic effect of introducing computer systems had not been used to obtain some reorganization gains.

128

6.9.2. Adaptation

The fact that all three systems originated with user management meant that distribution managers were closely involved in the design processes and in this way were able to create systems that met their needs and which they fully understood. Considerable efforts were made to interest the clerks in systems design. A formal consultation structure was created for the development of the home sales system and this proved successful, for the clerks approved of the system. Consultation appears to have been less successful in the export section perhaps because of the greater complexity of the design problems. No formal consultative structure was associated with the batch system but efforts were made to involve the clerks in the introduction of the on-line system. The systems designer saw the clerks as passive and unable to contribute ideas but the export area distribution manager made the valid point that it was difficult to discuss a system when it did not exist. People not used to computer systems could not make constructive comments until they actually saw the system. There was always a gap between technical expertise and user understanding. In the author's experience clerks can play a role in the design of their own systems but they have to be given the skills to do this. Consultation without these skills encounters the problems just described.

To summarize, at management level adaptation to both of the systems was good. In the home sales section, where there were few implementation problems, clerical adaptation was also good. Work difficulties in the export section at the time when both the batch and on-line systems were being introduced caused some resistance, and the inflexibility of the batch system caused frustration to a highly skilled group of clerks who were used to working in a flexible and responsible way. It may be that the systems designers did not fully understand the complexity of the export order clerks' job when they designed this system.

6.9.3. Integration

The integration of the technical and human systems was, as we have seen, excellent in the home sales section and contributed to a better relationship with the customer environment through the speeding up of order processing. Integration was less successful in the export section and management attempted to remedy difficulties there by reviewing the organization of work. Efforts were made to increase the interest of work at the lower levels of the clerical hierarchy by combining the jobs of document disposal clerk and invoice clerk but these changes did not appear to be popular. In the 1978 survey, whereas the order clerks described the on-line system with approval, other clerks spoke of the computer causing them increased pressure and frustration. A solution here might be to encourage the clerks to develop their own form of work organization, so that they, rather than management, are the change agents.

6.9.4. Pattern Maintenance

Once integration has been successfully achieved then a work system has to be kept in a state of balance, providing an efficient and satisfying environment for the staff working in it, yet able to respond easily to challenge. At the time the post-change survey was carried out the home sales section had achieved this state of balance, the export department was still working towards it.

7. The Industrial Firms:
B. Asbestos Ltd.

Asbestos Ltd. is a subsidiary of a consortium of companies. The consortium was formed in 1920 as a result of an amalgamation of four companies in the asbestos, chemical and insulation industries. One of its principal objectives was to "manufacture, deal in, erect and supply materials, substances and appliances for affording insulation or protection from heat, light, electricity, sound, blows, shocks, vibrations, air, water, fluids, gases, emanations and rays". The basic raw material which would enable the company to do this was asbestos in association with some allied products.

The asbestos part of the consortium emerged from the Lancashire cotton industry when the sons of one of the Rochdale millowners began to manufacture a revolutionary type of lubricated cotton packing ring for keeping piston rod and valve spindle stuffing boxes steam tight. This activity later developed into the production of asbestos yarn and cloth to be used as packing for high temperature engineering applications. Supplies of asbestos were found and mined in Canada. The firm then began making a wide range of asbestos products and these eventually replaced the original range of cotton goods made at the Rochdale Mill. In 1913 the firm opened a branch factory on the new Trafford Park, Manchester, Industrial Estate. This was the beginning of what would eventually become Asbestos Ltd., the largest single manufacturing unit within the consortium, based not upon asbestos textiles but upon asbestos cement building materials. Today building and construction materials account for nearly a third of the consortium's turnover. The principal outlets for Asbestos Ltd. products are in industrial and agricultural buildings. These can be roofed and clad in asbestos-cement sheeting. Given the present controversy over asbestos and health, the firm is increasingly transferring its activities from asbestos to substitution materials.

7.1. THE BUILDING AND INSULATION DEPARTMENT: THE BATCH COMPUTER SYSTEM

Computers have been used extensively in many parts of Asbestos Ltd. since the 1960s. The department selected for this study dealt with the processing of orders for asbestos sheeting and moulded rainwater goods – items which individually were small and inexpensive but which were sold in large quantities to the building industry. This department had installed a batch processing system in 1970 and in 1973 the author was asked by the company if she would evaluate the human consequences of this system for the clerks in the building and insulation (B&I) department as a guide to the design of the next system. In 1974 an on-line system incorporating visual display units was introduced. The author gave advice on the design of the human part of this system, that is, on the organization of work and the allocation of group and individual task responsibilities, and was later able to evaluate the impact of this second system on job satisfaction.

The introduction of the batch processing system was not easy. It was the first sizeable system introduced into Asbestos Ltd. and it required new kinds of discipline from the clerks in the B&I department. It was introduced at a time when business was not very good and employees were being made redundant. In consequence the computer became associated with redundancy. In 1972 the management services group and the senior manager of the B&I department began to consider a new, improved form of system. There was a desire to reduce the cost of processing orders, to bring clerks into closer contact with the computer through direct interaction and to develop a system which would better fit their needs. With the batch system orders had to be coded from a manual to a computer format, batched, and then sent to the computer centre where they were punched ready for input to the computer. Orders were processed overnight and this meant a twenty-four hour delay before B&I received information. If orders were rejected by the computer then they had to be resubmitted the following morning which caused an additional delay.

An on-line interactive system would provide a faster means for processing orders and would eliminate the coding functions, thus saving staff. Instead of requiring a group of special clerks to handle data control, a single clerk would be able to input the order information directly to the computer via the visual display units and order confirmations could be produced on a line printer almost simultaneously.

The basic tasks of the B&I department were as follows:

1. The processing of customers' orders, either manually or via the computer.
2. Ensuring satisfactory service for all orders received.
3. Disseminating the statistics required for processing an order and making alterations for cancelling orders as customers revise their needs.
4. Maintaining and updating the customer files and records in the department.

With the batch computer system the work of the department was divided into four main sections with supervisors in charge of each section. These were:

(1) *The territory section.* The main function of this section was to answer customers' queries and to do this it was sub-divided into two regions: the north and the midlands. Although some queries arrived by mail, the vast majority were telephone requests for information. New orders and alteration or cancellation of existing orders were received by telephone or telex and these had to be recorded and passed on to the relevant section in the department for processing. The territory clerks' job could be pressurized or boring depending on the pace of the work and the nature of the queries. The territory section had a staff of twelve clerks.

(2) *The order allocation section.* The main task of this section was the allocation of orders to factories producing the required goods and the assessment of delivery dates. There were a number of small factories associated with Asbestos Ltd., and B&I acted as a centralized order processing department linking customers to these. Queries concerning allocations or delivery dates were handled by this section. There was a staff of three clerks.

(3) *The order preparation section.* This section coded orders which were to be processed by computer and copied the few orders which were still processed manually. Clerks also dealt with customer queries concerning materials and quantities. There was a staff of ten.

(4) *The data control section.* This section acted as the interface between the sales order department and the EDP department. The clerks recorded and batched those orders which were to be processed by computer. They also recoded altered orders and received computer print-out. This included invoices, acknowledgements of orders and error reports. There was a staff of five.

There were a number of constraints affecting the processing of work in the B&I department. Little work could be done in the morning until either the mail or the computer print-out from the overnight run was received. Orders to be processed by computer had to be ready by 2.00 p.m. and if this deadline was not met they were delayed until the following day.

Departmental work flow was as follows:

(1) Orders were sorted into regions (north and midlands) and given to the section leaders of the territory section handling that region. Postal orders were checked against telephone orders to establish if they were confirmations of orders already received or new orders. The orders were duplicated and one copy was kept by the territory section, the other going to the order allocation section.

(2) Orders passed to the order allocation section for allocation to factories and determination of delivery dates. In allocating orders consideration is given to the proximity of the customer to the factory, the need to avoid split deliveries (orders being sent in two parts) and the ability of the factory to meet a required delivery date.

(3) After order allocation, orders and their debits were recorded by a clerk in the data control section before going to the order preparation section (4) (5). Both manual and computer orders were copied onto draft sales order forms by the order preparation clerks.

(6) Completed draft sales orders were batched and recorded before being sent to the computer centre (7).

(A) Each morning a print-out of the previous day's input was received from EDP by the data control section. This print-out consisted of: invoices of despatches made, acknowledgements of new orders and alterations of orders (ATOs), error reports and other documents such as product or price lists which had been requested.

(B) Invoices and acknowledgements were sorted into those for customers, sales representatives, branch offices and customer files. Those for customer files were

Figure 1. Asbestos Ltd. Work flow of building and insulation department

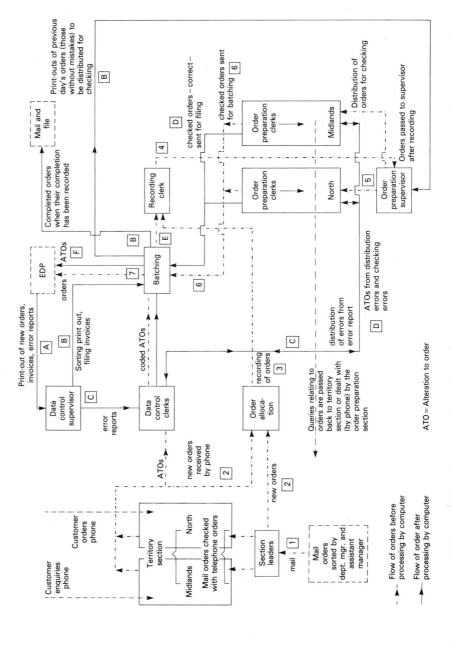

linked with new orders or ATOs. The batching clerk checked invoices against customer files for completion.

(C) The error report was examined on the data control section; errors in coding were returned to the order preparation clerk.

(D) The order clerk prepared a manual ATO to rectify an error and passed it, with the customer file, to the data control clerks for coding.

(E) (F) Coded alterations were handed back to the batching clerk to be batched and recorded before being sent to EDP.

Simultaneously with this flow of order processing the clerks on the territory sections received telephone orders and handled queries.

The task structure and work flow associated with the on-line system will be described later in this chapter.

7.2. THE IMPACT OF THE BATCH COMPUTER SYSTEM ON JOB SATISFACTION

The introduction of the batch computer system effectively split the B&I department into two main groups of staff: territory clerks and order preparation clerks. Territory clerks had varied and challenging work because of their responsibility for handling customer queries. They had considerable work freedom and opportunities for using initiative and judgement, but they worked under a great deal of pressure as dissatisfied customers telephoned to know why their goods had not arrived or why they had been sent the wrong goods. They also had to respond to customers who wanted goods urgently or who wished to change an order they had already submitted.

In contrast, the order preparation clerks had routine but unstressful jobs. Eighty percent of their time was spent in coding draft orders for the computer; they had little work variety and almost no opportunities for using discretion. This was a solitary job which required few contacts with other members of their group and fewer still with the rest of the department. Table 1 compares the two job structures.

A job satisfaction survey was carried out in the department using the framework for analysis described earlier but as the questions were worded differently the data cannot be directly compared with the second survey carried out after the on-line system was installed. Although the territory clerks had considerable work variety, they indicated that they were too tied to the telephone and would like to have a broader range of tasks. In answer to a number of questions on variety and routine around 75% of the order preparation clerks said that they would like to have more variety and less routine.

The message for any future computer system that was designed with job satisfaction as an objective was to reduce the polarization of work by freeing the territory clerks from the pressure of constantly having to respond to customer queries, while providing the data preparation clerks with more interesting and challenging jobs.

135

Table 1. Asbestos Ltd. Job structure comparisons

CONTROL LEVELS	LEVEL 1			OPERATIONAL LEVEL			LEVEL 2	LEVEL 3	LEVEL 4	LEVEL 5
								HIGH CONTROL LEVELS		
	Variety	Means choice	Knowledge choice	Job identity	Give sense of team work	Job relationships	Goal clarity	Anti-oscillation	Optimisa-tion / Develop-ment	Overall control
									Creativity / Autonomy	Key task
	Give sense of personal control	Give feeling of achievement	Give sense of making important contribution			Give sense of confidence	Problem prevention	Efficiency improvement		

Definition of skills:
A. Communicating in writing
B. Communicating verbally
C. Arithmetical skills
D. Machine operating skills
E. Checking/monitoring/correcting
F. Problem solving
G. Coordinating
H. Supervising

These are the day to day operating tasks

Column sub-headings (left to right):
- Reduce boredom — No. of tasks, No. of skills, of methods, of work sequence
- Use of judgement
- Use of initiative, Clearly defined start to job, Clearly defined end to job, Uninterrupted task sequence
- Long task cycle 20+ minutes, Visible contribution to product or service
- Works as member of group
- Considerable interaction required, Clear work objectives
- Objectives not too difficult easy or too difficult, Can requisition required resources
- Can correct errors, solve problems
- Coordinates own work activities, Coordinates group work activities
- Can improve methods, Can improve product or service, Individual free from supervisory control, Group free from supervisory control
- Problem solving, Coding

	Large / Small	No. of tasks									
Territory clerk	Large	A/B/E F/G	5	√ ...							Problem solving
Order prep. clerk	Small	E	1	√ ...							Coding

7.3. THE ON-LINE COMPUTER SYSTEM

An understanding of the philosophy of the design processes associated with the on-line system in the B&I department will be helped by examining the values of the different groups involved in these. The survey of values was carried out after the implementation of the on-line system. An important objective of this had been to increase job satisfaction and produce a computer-based work system that was both technically efficient and satisfactory in human terms. The motivation for this approach had come primarily from the systems design group who were well aware of the human deficiencies of the earlier batch system.

7.4. ORGANIZATIONAL VALUES

All three groups, clerks, systems designers and managers – the departmental manager and his senior manager – saw the current values of the firm as on the left hand side of the scale; with high priority placed on discipline, standardized methods and tightly structured jobs. The main difference of opinion was on whether the firm believed in shared values or tolerated a multiplicity of different values, with the clerks believing that the latter was the case. Preferred values for the clerks and systems designers were mainly in the centre or on the right hand side of the scale, although both groups wanted shared values. The two managers were less theory Y-oriented and preferred discipline and standardized methods and procedures. These scales suggest that there were few differences of opinion on organizational values between the systems designers and the clerks, but some disagreement between the managers and the clerks.

The systems designers' view was that the company was in the process of moving itself from the left hand to the right hand side of the scale and that this was causing some value conflicts, with parts of the firm still organized around discipline and structure and other parts becoming more flexible and democratic. These new values were shown in improved communication and consultation and in a less authoritarian style of management. They saw their own design task as influenced by this shift in values and their role as trying to produce a structured technical and organizational framework which, at the same time, permitted task flexibility.

The managers, with their emphasis on discipline and standardized methods, explained that staff should discipline themselves and that discipline should not have to be imposed by management. Standardized methods provided a consistent framework, but within this staff could be allowed to use their own discretion. They also believed that the aim of shared values was unrealistic. The senior manager commented:

> We have to accept a multiplicity of values but these should operate within a framework which provides an explanation of company objectives, discussion of these and a certain amount of commitment. This does not mean that staff must necessarily agree with the objectives of the firm, but after reasoned explanations they should accept and work to them.

Figure 2. Perceived organizational values (modes)

Place a tick where you think this firm fits on the scale below.

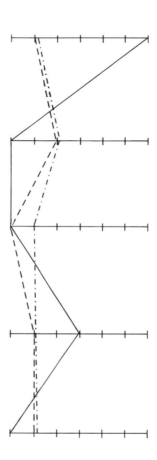

It has a strong belief in the importance of shared values – that employees should agree with its objectives and the way it carries out its operations

It likes to have disciplined employees who will put the firm's/office's interests first and willingly accept orders and instructions

It prefers to use standardized methods and procedures whenever possible as it believes that this assists efficiency in this kind of firm/office

It places a great deal of emphasis on efficiency and high production and less on personal qualities such as friendliness, trustworthiness, cooperation etc.

It feels the need to organize work activities into tightly structured jobs which are clearly defined and do not permit a great deal of individual discretion

A multiplicity of different values exists and is tolerated here. It is quite acceptable for employees to hold very different views on what the firm should be doing and how it should be doing it

It does not believe in very strict discipline and tries to provide a situation in which people can pursue their own interests

It likes employees to work out their own methods for doing things and to use their own judgement when taking decisions

It places a great deal of emphasis on personal qualities such as friendliness, trustworthiness, cooperation etc. and less on efficiency and high production

It tries to organize work in such a way that employees have loosely defined and structured jobs which permit a great deal of individual discretion

```
——————————  N = 22   Clerks
— — — — — —  N =  2   Managers
— · — · — · —  N =  2   Systems designers
```

138

Figure 3. Desired organizational values (modes)

Please tick where you *would like* this firm to fit on the scale below.

It should have a strong belief in the importance of shared values – that employees should agree with its objectives and the way it carries out its operations

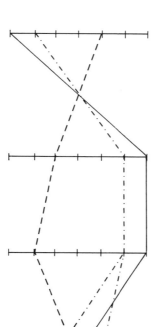

A multiplicity of different values should exist and be tolerated here. It should be quite acceptable for employees to hold very different views on what the firm should be doing and how it should be doing it

It should try to create disciplined employees who will put the firm's/office's interests first and willingly accept orders and instructions

It should have an easy-going kind of discipline and try to provide a situation in which people can pursue their own interests

It should try to use standardized methods and procedures whenever possible as this would assist efficiency in this kind of firm/office

It should encourage employees to work out their own methods for doing things and use their own judgement when taking decisions

It should place a great deal of emphasis on efficiency and high production and less on personal qualities such as friendliness, trustworthiness, cooperation etc.

It should place a great deal of emphasis on personal qualities such as friendliness, trustworthiness, cooperation etc. and less on efficiency and high production

It should try to organize work activities into tightly structured jobs which are clearly defined and do not permit a great deal of individual discretion

It should try to organize in such a way that employees have loosely defined and structured jobs which permit a great deal of individual discretion

————————— N = 22   Clerks
– – – – – –   N = 2   Managers
– · – · – · –   N = 2   Systems designers

139

Figure 4. Organizational models (modes)

Please tick on the scale below what *you* believe to be the best form of department structure for clerical staff in general.

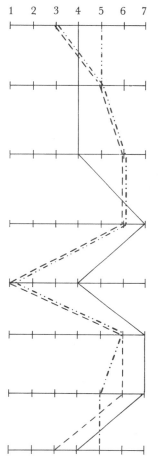

	1	2	3	4	5	6	7	

Jobs should be clearly defined, structured and stable — Jobs should be flexible and permit group problem solving

There should be a clear hierarchy of authority with the man at the top carrying ultimate responsibility for all aspects of work — There should be a delegation of authority and responsibility to those doing the job regardless of formal title and status

The most important motivators should be financial, e.g., high earnings and cash bonuses — The most important motivators should be non-financial, e.g., work challenge, opportunity for team work

Jobs should be carefully defined by O&M department, management services or supervision and adhered to — The development of job methods should be left to the group and individual doing the job

Targets should be set by supervision and monitored by supervision — Targets should be left to the employee groups to set and monitor

Groups and individuals should be given the specific information they need to do the job but no more — Everyone should have access to *all* information which they regard as relevant to their work

Decisions on what is to be done and how it is to be done should be left entirely to management — Decisions should be arrived at through group discussions involving all employees

There should be close supervision, tight controls and well maintained discipline — There should be loose supervision, few controls and a reliance on employee self-discipline

```
————————  N = 22   Clerks
— — — — —  N =  2   Managers
·—··—··—··  N =     Systems designers
```

140

Figure 5. Models of man (modes)

Please tick on the scale below how you see the general characteristics of the majority of staff in this department.

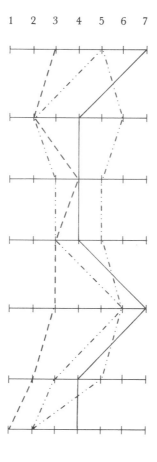

They	They
Work best on simple, routine work that makes few demands of them	Respond well to varied, challenging work requiring knowledge and skill
They are	
Not too concerned about having social contact at work	Regard opportunities for social contact at work as important
They	
Work best if time and quality targets are set for them by supervision	Are able to set their own time and quality targets
Work best if their output and quality standards are clearly monitored by supervision	Could be given complete control over outputs and quality standards
Like to be told what to do next and how to do it	Can organize the sequence of their work and choose the best methods themselves
Do not want to use a great deal of initiative or take decisions	Like, and are competent to use initiative and take decisions
Work best on jobs with a short task cycle	Able to carry out complex jobs which have a long time span between start and finish

——————— N = 22   Clerks
– – – – – – N = 2   Managers
· — · · — · · — · · N = 2   Systems designers

141

The clerks' profile of the best form of departmental structure for clerical staff in general was consistently in the centre or down the right hand side of the scale, with flexibility, delegation of authority, group decision taking and few controls. Systems designers and managers were also in the centre or on the right hand side of the scale although they came out strongly for targets to be set and monitored by supervision. There was a wider divergence of opinion on the general characteristics of the staff of the B&I department. The clerks were either in the centre of the scale or on the right hand side, seeing themselves as responding well to varied, challenging work and able to organize their own work. Their managers saw them as on the left hand side of the scale, at their best in structured situations where they were not given much discretion. The two systems designers were divided on four of the seven statements, one seeing the clerks as more flexible and responsible than the other.

The four scales suggest that whereas the systems designers broadly see the clerks and interpret their needs in the same way as the clerks themselves, the view of the managers does not fit so closely to those of the clerks. This might indicate that the systems designers would provide the kind of computer system that fitted the clerks' needs but would have to persuade line management of the validity of their approach.

7.5. WORK VALUES

The factors systems designers and managers saw as producing efficiency, happiness and a sense of fairness and justice are set out in Figure 6.

Figure 6. Work values (systems designers' and managers' view)

EFFICIENCY

Role expectations
(what is expected of staff)

Cognitive	**Cathectic**	**Moral**
Willingness to learn	Loyalty	
Concentration	Commitment	
Awareness of importance of	Enthusiasm	
one's own role	Motivation	
	Job satisfaction	

Role requirements
(what the organization must provide)

Cognitive	**Cathectic**	**Moral**
Clear objectives	Management understanding	
Required resources	of individual needs	
Training	Opportunities for	
Good decisions	involvement	
Good operating systems		
Flexible work organization		
Clear job requirements		
Performance monitoring		

(Figure 6 continued on page 143)

(Figure 6 continued from page 142)

HAPPINESS

Role expectations
(what is expected of staff)

Cognitive	Cathectic	Moral
Acceptance of job definition	Feeling of accomplish- ment	

Role requirements
(what the organization must provide)

Cognitive	Cathectic	Moral
Achievement of department objectives	Opportunities for personal development	Good pay
Information flow	Good social mix of people	Good promotion
		Job security
		Consultation

FAIRNESS AND JUSTICE

Role expectations
(what is expected of staff)

Cognitive	Cathectic	Moral
—	—	—

Role requirements
(what the organization must provide)

Cognitive	Cathectic	Moral
An understanding of policies and objectives		Effective consultation
An explanation of decisions		Fair rules and procedures
		Effective appeals procedure
		Equitable pay scale
		Fair promotion opportunities
		Trade union

The answers to the question on what produces efficiency in an organization show that Asbestos Ltd. has strong expectations of its staff. Staff should possess certain intellectual traits such as willingness to learn and psychological disposi- tions such as loyalty and commitment. In return the organization must provide a rational and well-organized work situation and an appropriate emotional climate. This emphasis on the attitudes of staff to their work suggests a managerial interest in producing an environment which would stimulate the desired attitudes. Hap- piness is seen as having cognitive, cathectic and moral components and fairness and justice as related to equity and procedures for assisting this.

Job satisfaction was defined in the following way by the systems designers:

—Going home at the end of the day feeling you have contributed. Being happy to talk about what you have done.

—Wanting to go to work because you like going there. You would go even without pay if necessary.

and by managers as,

—A feeling of accomplishment.
—Knowing what my job is and being trained to do it. Knowing how it contributes to the "whole" and seeing the eventual result. Having responsibility and authority, and pleasant working conditions.

The systems designers' values and philosophy concerning their own work were determined by asking them to describe their approach to systems design, and the design principles which they found most useful. Each ranked the activities he included in systems design in the following way.

Systems designer A	Systems designer B
Define the needs of the business	Define the system objectives
Ensure good project control	Collect and analyse data
Identify the needs of user management and staff	Formulate options, define constraints
	Choose option, discuss with user
Develop good quality documentation	Specify system
Ensure good selling	Get final user agreement
Ensure good testing	Define and agree controls
Ensure good training	

Their design principles were based on the belief that it was more important to design a business system than a computer system. Priorities must be set at an early stage, user staff involved in the design processes, and the design of the technical part of the system should proceed in step with the design of methods and tasks. They said that their interest in involving the user in the design of his own system had been stimulated by their contact with the Manchester Business School. They had been influenced to place the different design alternatives before the user and let him make the final choice.

7.6. GOAL SETTING AND ATTAINMENT FOR THE ON-LINE SYSTEM

The task of the B&I department was (1) to convert orders from the customer into orders which could be processed by computer and passed to the factory, (2) to decide which factory should have the order and allocate a delivery date and (3) to progress orders, once issued, until despatched. The senior manager and the systems designers believed that an on-line system would improve (1) by providing a faster and less clerically heavy means for processing orders for the computer, in this way speeding up and simplifying the order processing activity. They hoped that an on-line system would also save staff, improve customer service and increase the availability of information.

The systems designers also saw the introduction of an on-line system as providing them with the opportunity for some exciting and challenging work and with the flexibility to try new forms of work organization. The user manager was convinced by the systems designers that a system using VDUs would save his office a great deal of time and money.

The technical objectives set for the system were interesting and unusual in that they related to the attainment of human objectives. The systems designers wanted to design a system that would remove the barriers between the clerks and the computer. They believed that the numbers and codes of the batch system did create such a barrier and that the use of VDUs would eliminate it. They also wanted to create a technical design that would not constrain the meeting of business and human needs. Technical objectives were therefore seen as closely interlocked with business and human goals.

Business objectives were primarily saving time and reducing costs, the one assisting the other. Human objectives were fitting the system to the needs of the clerks; providing them with the means to be more efficient; giving them more interesting and satisfying work and removing some of the controls which were associated with the batch system. These objectives met the needs of departmental and senior management and so there was no conflict of objectives.

Investigations into the viability of an on-line system began in 1972 and initially were a combined effort between the senior manager of B&I and one of the systems designers, with advice and assistance being provided by the computer manufacturer. A feasibility report was prepared by a working party consisting of three user representatives and one of the systems designers and this set out the benefits of a change to an on-line interactive system. A cost-benefit analysis was also made at this time.

7.6.1. Design Alternatives

The major influence on the choice of technical solution was a desire to use VDUs and the technical design alternatives considered were related to the best way of using VDUs. The technical part of the system was eventually framed as shown in Table 2 with the social and business advantages and disadvantages of each option clearly identified.

An alternative approach would have been to move to a decentralized computer system based on mini-computers. This was rejected as not fitting into the long-term plans of the department and also because the firm had no technical expertise with this kind of system.

Human design alternatives were worked out in collaboration with the author and her research group at the Manchester Business School. These are listed in Table 3 together with their social and technical advantages and disadvantages.

The systems designers recommended to the user that alternative 2 should be implemented, with a specialist group of VDU operators also accepting responsibility for data control. Their argument was that such an arrangement would lead to more work interest for the data preparation clerks who would now become VDU operators with some clerical functions. At the same time there would be an efficient, specialist group to operate the VDUs. But by this stage of the design process user clerks, who through working parties and discussion groups had become very involved in design activities, had made up their own minds on the most desirable form of work structure and they opted for alternative 3. This alter-

Table 2. Technical system alternatives

VDU input and output	Social/business advantages	Social/business disadvantages
Input new orders amendments enquiries	1. Provides opportunity to change present work organization	1. Staff do not want to use keyboards (a job seen as female)
Update on-line — order file batch — product files customer files	2. Staff will have more control over paperwork (immediate error correction and shorter cycle time)	2. This is another organizational change requiring staff adaptation
Output on-line — errors confirmation enquiries checking information	3. There will be job completeness. A job will be finished in a day	3. The response time of the VDUs will tend to pace work.
print out — queries acknowledgements lists of orders	4. More individual control over work pace	There will be hardware dependence with problems of breakdown
	5. Better close down time of computer (5 p.m. not 2 p.m.)	

146

Table 3. Human design alternatives

Organizational alternatives	Advantages	Disadvantages
(1) Similar to pre-change with: 1. VDU section; 2. Data control; 3. Territory sections (T) data control VDU T1 T2 T3	*Social* Minimum disturbance to existing work organization. Necessitates only the addition of a VDU section. *Technical/efficiency* Special VDU section can reach high level of performance.	*Social* Retains problems of existing organization. Territory clerks under pressure, other staff doing boring, routine work. *Technical/efficiency* System vulnerable to absenteeism and labour turnover.
(2) Specialist VDU section operators operate VDUs and do data control. Territory sections very similar to present organization. VDU + data control T1 T2 T3	*Social* Work is more varied for VDU operators. *Technical/efficiency* Special VDU section can reach high level of performance.	*Social* Territory clerks still working under pressure. *Technical/efficiency* System vulnerable to absenteeism and labour turnover.
(3) No specialist VDU section. Geographic split into sections according to territories. Staff in each section able to do all the jobs, with rotation. G1 G2 G3 VDU + all other tasks	*Social* Very flexible system. All staff become multiskilled and able to do all jobs. Work is interesting for all staff. *Technical/efficiency* Special VDU section can reach high level of performance. Greater flexibility should lead to greater efficiency.	*Social* The department already has a grading system and staff might object to the removal of the job hierarchy. *Technical/efficiency* VDU input may be slower than with 1 and 2 because it is not handled by specialist operators.

(Table 3 continued on p. 148)

(Table 3. Human design alternatives, continued from p. 147)

Organizational alternatives	Advantages	Disadvantages
(4) No specialist VDU section. Geographic split into sections according to territories, as in 3. Hierarchy of jobs with staff specializing in particular jobs.	*Social* Somewhat flexible in that some clerks can operate VDUs. *Technical/efficiency* Special VDU section can reach high level of performance. Greater flexibility should lead to greater efficiency.	*Social* Some clerks have routine work. *Technical/efficiency* Vulnerable to absenteeism and labour turnover.
(5) Specialist VDU section. Territories section geographically organized and responsible for customer liaison and data control.	*Social* Relieves territory clerks of some pressure by making them responsible for other duties. *Technical/efficiency* Skilled VDU operators.	*Social* VDU operators may find work hard to tolerate. Territory clerks may not want to do data control. *Technical/efficiency* More territory clerks will be required if pressure is to be relieved.

148

native split the department into a number of multi-skilled autonomous groups, each group looking after a geographic area of the country and providing a total service to customers in that location.

In terms of the theory of job design [Cherns 1976; Davis 1972; 1975; and others] this is the best socio-technical solution. In efficiency terms it enables each group to take responsibility for its own errors and problems, and for coordinating its own work activities. In human terms it provides group members with the opportunity for becoming expert in all the jobs involved in serving a group of customers. Instead of promotion being based on length of service or a vacancy in the next grade, it is now associated with increased knowledge. Anyone who acquires the necessary skills can take on a higher grade job. The systems designers, delighted with the fact that the users were knowledgeable enough to make an informed choice, willingly accepted the proposed solution. This system was therefore designed by what the author calls a "consultative" design approach, with users playing a major role in the design process but with the system still essentially designed by the technical specialists [Mumford 1978].

7.7. ADAPTATION

In Asbestos Ltd. the adaptation processes were influenced by the fact that the systems designers had approached the author for advice. In consequence, the results of the first job satisfaction attitude survey, used to test reaction to the batch computer system, were fed back to all staff in the B&I department in small groups. The purpose of this feed-back was threefold. First, to establish if the questionnaire had successfully identified the problems which staff regarded as important; second to gain an understanding of the reasons for these problems, and third to begin to get staff involved in thinking about new ways of organizing the department which would assist their solution. A steering committee was formed consisting of the systems designers, user management, and trade union and personnel department representatives. In addition four user committees were established to investigate the main design areas of the new system. There were also two working parties on implementation, a working party on the human aspects of the system and a project control committee. The working parties were drawn from experienced clerks in the B&I department, the project control committee from members of the development team.

Throughout this design process the technical and organizational parts of the system were designed simultaneously; higher priority was given to human than to technical needs and the diagnosticians of human needs were the clerks themselves. The view of the systems designers and the departmental manager was that this adaptation process had been extremely effective. Clerks in the department had become involved and interested in the new system, they were able to influence its design and development and they had a complete knowledge of how it would operate. This initial experiment with participative systems design stimulated the author to develop and systematize the approach and it has now been used in many other firms [Mumford, Land, Hawgood 1978].

149

7.8. INTEGRATION

Once the on-line system was implemented and had become operational the author spent some time in the B&I department observing the new organization of work and getting the clerks to complete a second questionnaire. Informal discussion gave her the impression that the on-line computer system was regarded as greatly superior to the old batch system and that all staff liked the new departmental structure in which all clerks were now territory clerks with responsibility for a total customer service including operating the VDUs. The only reservation about the system was that the poor level of trade in 1975 meant that there was no opportunity to test the system out under normal trading conditions.

Morale in the department appeared to be high and the new form of work organization gave individual clerks a great deal of work autonomy which they enjoyed. The ability to tackle customer problems was now related to knowledge rather than to function. This meant that new clerks had an excellent opportunity to learn on the job and were able to take on more decision making responsibilities as their knowledge of how to handle problems increased.

7.8.1. The Impact of the Computer on the Structure of Work

Figure 7 shows the job of the territory clerk with the on-line system, Table 4 the job structures of the two main groups of clerks in the department with the batch system, and the job structure of the new multi-skilled clerk. It can be seen that the job of territory clerk has expanded and more tasks provide relief from answering telephone queries. The routine job roles have now been eliminated. Figure 8 shows the work flow associated with the on-line system.

7.8.2. The Clerks' Perception of Integration

In order to ensure comparability with the other organizations participating in this research, job satisfaction data in Table 5 are presented in negative rather than positive form. That is, they show the percentage of staff who were unhappy with different aspects of their work. When examining these results it must be remembered that because of the small size of the sample 9% is only two clerks.

The results do not identify any major problem areas. The autonomous group structure has not removed work dependency and mistakes made by one member of a group affect his colleagues, but only 23% of staff (5 clerks) would like less dependency. All clerks believe that they can make decisions and use judgement although 45% would like even more opportunities to do this, and 14% believe that the on-line system has increased their opportunities for decision taking.

There was considerable agreement on the kinds of decision that the clerks most enjoyed taking and they spoke of decisions which helped the customer. Only 19% (4 clerks) thought that there was too little challenge in their work and 41% said that the on-line computer system, and associated reorganization of work, had provided more opportunities for challenge. It also helped them to meet deadlines.

150

Figure 7. Territory clerk

Looks after group of customers. Has 35 active ones. Autonomous job. Indexing and sending documents to mail and file takes longest part of the day. Solving customer problems is most complex part of job (⅓ of time).

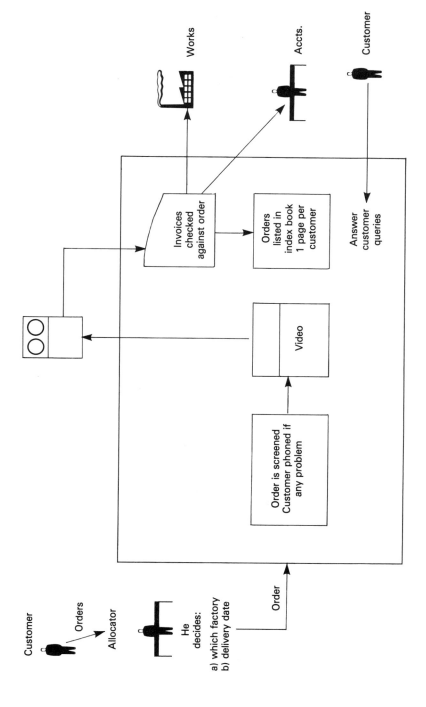

151

Table 4. Asbestos Ltd. Job structure comparisons

CONTROL LEVELS																									
																			HIGH CONTROL LEVELS						
		LEVEL 1 OPERATIONAL LEVEL																	LEVEL 2		LEVEL 3		LEVEL 4		LEVEL 5
		Variety			Means choice		Knowledge choice		Job identity						Job relationships		Goal clarity		Anti-oscillation		Optimisation		Development		Overall control
Definition of skills: A. Communicating in writing B. Communicating verbally C. Arithmetical skills D. Machine operating skills E. Checking/monitoring/correcting F. Problem solving G. Coordinating H. Supervising	These are the day to day operating tasks	No. of tasks	No. of skills	of methods	of work sequence	Use of judgement	Use of initiative	Clearly defined start to job	Clearly defined end to job	Uninterrupted task sequence	Long task cycle 20+ minutes	Visible contribution to product or service	Works as member of group	Considerable interaction required	Clear work objectives	Objectives not too easy or too difficult	Can requisition required resources	Can correct errors, solve problems	Coordinates own work activities	Coordinates group work activities	Can improve methods	Can improve product or service	Individual free from supervisory control	Group free from supervisory control	
		Reduce boredom		Give sense of personal control		Give feeling of achievement		Give sense of making important contribution					Give sense of team work		Give sense of confidence		Problem prevention		Efficiency improvement		Creativity		Autonomy		Key task
Batch system Territory clerk	√	Large	5 A/B/E F/G	√		√	√	√	√	√	with some queries	√	√	√	√	√	√	√	√						Problem solving
Order prep. clerk	√	Small	1 E					√	√	√					√										Checking
On-line system New territory clerk	√	Large	6 A/B/D E/F/G	√	√	√	√	√	√	√	√	√	√	√	√	√	√	√	√	√	√				Order handling Problem solving

All activities in the diagram below are carried out by a group of clerks for a specific group of customers. Orders are put into the computer via the VDUs. Information on stock positions and customer records can be obtained via the VDUs.

153

Table 5. Asbestos Ltd. Opinion on different aspects of work
(in percentages)

Clerks N = 22

Category	Aspect	Type	Statement	%	Comparison with batch system
VARIETY	Routine and challenge	FACT	Work is usually or always routine	27	
		PREF	There is too little challenge	18	But more challenge than before: 41
		PREF	There is too much routine	32	
		FACT	Almost everything has rules or set procedures	—	
DISCRETION	Means choice	FACT	Can hardly ever choose our work methods	9	
		FACT	Can hardly ever choose work sequence	9	
		FACT	Can hardly ever choose next task	27	
		PREF	Would like more choice of work methods	14	
		PREF	Would like more choice of work sequence	14	
		PREF	Would like more choice of next task	—	
		PREF	Would like more opportunities to make decisions, use judgement	45	But I take more decisions now: 14
		FACT	Can hardly ever make decisions, use judgement	—	
	Knowledge choice	FACT	Has very little or little chance to see a piece of work through from start to finish	18	
		PREF	Would like more chance to see a piece of work through from start to finish	50	
JOB IDENTITY	Task cycle and integration	FACT	Inefficient work by others can almost always or usually hinder me	64	
		PREF	Would like more chance to do this	64	
JOB RELATIONS	Work dependency	FACT	Inefficient work by me can almost always or usually hinder others	23	But I am less dependent now: 14
		FACT	I would like there to be less work dependency	32	
		PREF	There is too little or barely enough time to do work	32	
GOALS	Deadlines	FACT			But deadlines are easier to meet: 36

Table 6. Asbestos Ltd. Positive and negative feelings about work (in percentages)

Negative feelings about work

	FACT	FACT	FACT	FACT	FACT
Time drags ⅓ of the day or more	23				
My present job is far from my ideal job 1/2		27			
I am very little involved in my job			14		
Taking the job as a whole I do not enjoy it				18	
I hardly ever get a sense of achievement					14

Clerks N = 22

Comparison with batch system

Time drags on for me more often	18
I enjoy it less now	18
I get less sense of achievement now	23

Positive feelings about work

	FACT	FACT	FACT	PREF	FACT
Time never seems to drag	41				
My present job is close to my ideal job 5/6		59			
I am strongly involved in my job			41		
Taking job as a whole I enjoy it very much				50	
I would like to get sense of achievement more often					64
I get sense of achievement almost every day					45

Time drags for me less often	27
I enjoy it more now	45
I get more sense of achievement now	32

155

General measures of job satisfaction show a largely satisfied work force. 45% said that they got a feeling of achievement every day and 32% said that the new system had enabled them to get more sense of achievement. 19 clerks provided examples of what gave them the most sense of achievement and their answers can be categorized as "satisfying customers" (9), "solving problems" (4) and "efficiency and a high standard of work" (6). 59% also said that their present job was close to their ideal job.

Despite these positive replies there were some clerks who did not like the computer, although their objections appeared to be philosophical.

> —Automatic processing deadens the senses, it removes all sense of initiative.
> —We become a slave to a machine. Everything seems to revolve around what man can do for the computer, rather than what the computer can do for us.

From the clerks' point of view the on-line computer system was an example of good socio-technical integration. The design approach had welded together a department that was previously not well integrated, but had a functional organization that both compartmentalized jobs and produced job polarization, with some jobs challenging yet stressful and others routine and monotonous. The new system used computer technology to assist this integration by enabling clerks to interact with the computer as part of their normal work activities. Information about customers could be obtained easily and rapidly via the VDUs although the system was not in real time and therefore was not completely up-to-date. The author was impressed by the number of elderly gentlemen anxious to show her their prowess in operating a VDU; they clearly took pride in having mastered this new skill.

7.8.3. Management's Perception of Integration

7.8.3.1. *The Fit with the Needs of the Clerks*

The system designers saw the main strengths of the clerks in lying in the value they placed on providing an excellent customer service. They had a great deal of loyalty and experience, identified with the task and the customer, and did their best to ensure that their work met the needs of the customer and assisted the image of the firm. The systems designers believed that their design philosophy had helped to exploit these strengths. Clerks were now able to provide an improved service to customers as a result of fewer errors and a faster enquiry facility. In their view the system had contributed to an increase in job satisfaction although they recognized that job satisfaction was dynamic rather than static and that the expectations of the clerks would alter over time. They hoped to be able to develop the system to meet these increased expectations. The departmental manager confirmed that the system had increased job satisfaction and said that morale was now higher than previously.

The system's contribution to the efficiency of the clerks was difficult to measure because the depressed level of business meant that it could not be tested under normal operational conditions. There were also some complaints from the department about technical faults such as VDU breakdowns, variable response times and static on the video screens. Although these deficits were at an acceptable level from the EDP department's viewpoint, they caused frustration to the user.

7.8.4.2. *The Computer's Contribution to Better Management Information and Control*

Both the systems designers and the B&I manager were agreed that the new system provided the user with more information. Information about customers, about whether a new order had been despatched and about the stock position could be called up on the VDU screens. The senior manager said that he was not getting any more information than before but that he now obtained this more easily and rapidly. Previously information had been up to a week out of date, now it was correct within 48 hours. Accuracy had been improved through the immediate feedback on input errors provided by the VDUs. In management's view the system had not led to more centralization as the B&I department already provided a centralized service, acting as an interface between the customer and the factory. It might eventually become decentralized with customer service centres based on factories, and the use of VDUs would assist such a change. It was hoped to move the updating of stocks onto a real-time basis.

7.9. MANAGEMENT'S EVALUATION OF THE SYSTEM

The systems designers' answer to the question "what aspects of the system best measure up to your idea of a successful computer system?", were:

1. It met the needs of the users.
2. It had returned the control of work to the user. The batch system had taken this away and given it to the computer.
3. It was easy to implement and in their view this was a measure of the success of a system.
4. The user liked the system and was able to use it efficiently.
5. The financial objectives had been attained and the system had led to financial gains.
6. It had proved technically effective and provided a basis for further developments.

They would have liked to design a system that could grow with the user, becoming increasingly difficult and demanding as the user became more skilled, but they recognized that it was not easy to design such a system. However they believed that the existing system had had an educational impact on the user, stimulating him to ask more of the computer. If users had not been involved in the design of the system they would not have been able to make these demands. The systems designers were pleased with the system and said that they would make few changes in their approach if they were designing it again.

The departmental manager said that the system had been highly successful in achieving the business objectives set for it. He was convinced that considerable benefits would be reaped once the department became busy again. There had also been some staff saving, although this had been achieved through natural wastage. The senior manager agreed with this assessment and praised the removal of the old conveyor belt form of work organization and the boundaries between functional sections.

7.9.1. The Computer System and Corporate Strategy

The senior manager did not think that the on-line system had contributed to improved business strategy. In his view business strategy meant marketing and the computer system did not provide the kind of information that would enable better marketing decisions to be taken. However, senior management were very aware of the potential of the computer to provide them with better information. This could assist the company to plan more effectively and respond to changes in its environment. For example, in recent years, because Asbestos Ltd. was linked to the construction industry, it had experienced extreme fluctuations in demand for its products. There had been a boom in 1973 and a slump in 1975. There had also been fluctuations in the availability of labour. Government wage freezes had caused employees to leave for more money and, at a time when demand for the product was going up, the ability to produce had been going down.

In his view the success of the on-line system was due to the skill and competence of the systems designers and to the importance they and he had placed on using the computer to improve job satisfaction. The introduction of VDUs had been of great benefit and the use of these would be extended throughout the company. There was now no resistance to the use of computers within Asbestos Ltd. although there were financial constraints and departments had to make a very good case for any money they required.

7.10. CONCLUSIONS ON THE ASBESTOS LTD. ON-LINE SYSTEM

7.10.1. Goal Setting and Goal Attainment

This was a successful system in that all the goals that had been set were successfully attained. These included technical, business and human goals, with the business and human goals being ranked higher than the technical goals. The system had led to cost savings through reducing the number of staff required to operate the B&I department, it had led to an improved customer service and it had increased the job satisfaction of staff through providing the opportunity for a redesign of the work structure of the department. The interaction of each set of goals with the others was recognized from the start of the design process with the result that the technical and organizational parts of the system had been designed simultaneously. The values of the systems designers had led them to place a high priority on human objectives and they were able to persuade a traditional man-

158

agement to accept a participative design approach and a form of work organization radically different from the existing one. Recognizing their own lack of experience in diagnosing and meeting human needs, they used the expertise of the
Manchester Business School to provide them with guidance in the design of the
human part of the system.

7.10.2. Adaptation

The adaptation processes were greatly assisted by the decision to use a participative approach to systems design. Clerks in the B&I department were involved in
the design processes from the beginning through group discussions and membership of committees and working parties set up to take responsibility for different
aspects of systems design and for the development of implementation strategies.
This policy enabled the clerks to become knowledgeable about the proposed
system and insightful into their own needs. In consequence, when the systems
designers suggested a particular form of work organization, the clerks were able
to offer a better alternative. The systems designers themselves went through an
adaptation process during the design of the system. Recognizing that they were
technical but not social experts and therefore lacking the experience to carry out
socio-technical design, they sought outside advice and acquired the knowledge
and experience to become effective social designers.

The departmental manager also adapted himself to a new situation. He was in
his late fifties, had worked all his life for the one firm and, initially, his value position was more theory X than Y. But, persuaded by the systems designers to accept a new, democratic design approach, he placed no obstacles in their way.
Once the system was implemented and seen to be successful his values changed
and he became an advocate of participation and of flexible work structures.

7.10.3. Integration

The B&I department was an example of a poorly integrated pre-change situation
that was transformed into a well integrated post-change situation. The original
batch system was slow and required a coding and checking operation before data
could be passed to the computer. This split the department into two separate
groups whose work was not integrated and whose task structures were very different. The on-line system eliminated the intermediate data preparation process and
enabled the department to be integrated into a number of multi-skilled groups all
providing a similar service to different groups of customers. Because the on-line
system assisted the efficiency of the clerks and also added to their job interest,
there was a good integration between the technical and social parts of the system.
In addition the computer assisted a better integration of the internal and external
environments. Customers could now be given information more quickly and their
problems solved more readily. This integration will be further improved when
there are VDU links between the B&I department and the factories.

159

7.10.3. Pattern Maintenance

This state of equilibrium will have to be managed if it is to continue into the future. The system designers are aware that job satisfaction needs alter and that employees' work expectations increase in times of prosperity, and they hope to be able to enhance the system to meet these new needs. Similarly the needs of customers change and in a highly competitive product market situation Asbestos Ltd. may require a computer system that responds even more speedily and effectively to customer demand; changing to a real-time mode will contribute to this.

8. The Government Department: The Inland Revenue Centre 1

Because the computer system was not completely implemented at the time of this study it was possible to compare computerized and manual departments.

Centre 1 is a modern office block in East Kilbride which is responsible for handling the tax returns and liabilities of the two million people who pay tax under P.A.Y.E. in Scotland. It has a staff of 1,800 and uses a computer to keep the basic records of tax payers and to carry out various routine calculations.

8.1. HISTORICAL DEVELOPMENTS IN THE INLAND REVENUE

Before the Inland Revenue began to use computers there were 200 tax offices in London, each dealing with a local tax area and employing a staff of about 40. In the late 1950s it became increasingly difficult to obtain clerical staff and so in 1958 the Board of the Inland Revenue took a decision to transfer P.A.Y.E. out of London and to create a number of larger London/provincial offices, each of which would handle the affairs of 150,000 to 200,000 tax payers and employ a staff of around 100. At this time the Board recognized that more modern methods for processing tax data were becoming available and so the first studies of computerization were also begun.

In the manual system all files were kept by reference to employers and these were listed in alphabetical order so that if a tax officer knew the name of a man's employer, he could find the employee's tax file very easily. Tax officers dealt with groups of employers and became familiar with their peculiarities. For example, they knew the different kinds of superannuation and holiday schemes which particular employers favoured. The first computer system aimed to replicate this manual filing system but it would have been very expensive to operate and so was not implemented. The problem was the movement of files. When people changed their employment their files had to be transferred to the new employer and while it was easy, if time consuming, to physically move a file from one part of a filing system to another, it would have been less easy to carry out this movement once files were a part of the computer system.

It was then decided to introduce static filing. That is to change the record when the employer changed but instead of the tax officer looking after groups of employers, he would take responsibility for the affairs of one group of tax payers irrespective of whom they were employed by. Tax payers were identified by their national insurance numbers and each tax officer dealt with around 4,000. This change to the national insurance number as identification simplified the change to computerized records.

Centre 1 at East Kilbride was intended to be the first of nine centralized tax offices located in different parts of Great Britain. It was selected as the pilot because in Scotland movement between jobs tends to be restricted to Scotland; therefore there was less chance of movement from computerized to non-computerized parts of the country. After Centre 1 was completed, building began on a second large centre in Bootle, Liverpool; this was dogged by industrial relations problems in the construction trades and before it was completed a government decision had been taken to abandon the P.A.Y.E. (pay as you earn) computer programme. Experience with Centre 1 showed that it was going to be very difficult to staff such large centres. They required the transfer of many staff and while East Kilbride was a new town and able to provide modern housing for civil servants, this would have been less true of the other centres.

It took two years, from 1967 to 1969, to transfer the records of two million employees and 93,000 employers to Centre 1 and to convert them into a form suitable for ADP.[1] In Centre 1 it had to be possible for the computer to examine each of the two million tax records daily, to establish which needed to be amended and to update about 43,000 of them. Tax records are now kept on magnetic tape, although the content of these records is little different to the previous manual system and they are still a summary of the tax payers' affairs. Because the ADP system had to be integrated with an existing manual system it had to be structured in exactly the same way. Centre 1 replaced 65 local tax offices in Scotland. Planning for the Centre started in 1963 and the systems work for the computer in 1964. All the planners were from the Inland Revenue although they were not necessarily all tax specialists.

Centre 1 employs a staff of 1,800 of whom 1,300 are in tax payer services, 200 in computer services, 200 in management services with responsibility for typing and other clerical services and 25 in personnel. Within tax payer services there are a number of allocation units in which staff deal directly with tax payers. Each allocation unit assumes responsibility for the files of tax payers within a section of the national insurance number range. Employers' files are held in the employers' unit and the staff there are responsible for disseminating to the allocation units all the information relevant to individual employees, for example, holiday pay and superannuation arrangements, agreed tips, etc. The introduction of ADP relieves the tax officer of the need to amend the tax payer's file; he now only has to notify the computer of the amendment and the computer will change the record and automatically notify the tax payer and his employer of changes in tax codes. The tax officer still has to correspond with tax payers and decide what allowances and reliefs are due to them. A major saving of time and staff is achieved through the computer's capability to deal with the heavy seasonal jobs that occur throughout the year. The computer division carries out the tasks of issuing returns and reminders, budget and annual coding. Some complex cases continue to go to the tax officer for manual treatment and these are notified on computer printouts. Other printouts concern "exceptions", when the computer rejects data or requires additional information. The computer therefore has removed much of the routine and repetitive work from the tax officer leaving him with cases that require judgement and decision.

1. The convention in the British Civil Service is to use the term ADP (automatic data processing) instead of EDP.

8.2. THE RESEARCH SITUATION

An introduction to the Inland Revenue was secured through the kind assistance of the Central Computer Agency of the Civil Service Department, and in January 1975, a meeting was held in London with tax officials who had had responsibility for the first design stages of the Scottish system. The original team had dispersed and it was suggested that the author should interview systems designers now working in London and Bootle in order to gain an understanding of the early part of the design process. In order to obtain comparisons between computerized and manual units it was decided that two allocation units should be examined, together with the employers' unit and a unit dealing with complex tax payers whose affairs were still handled manually. The employers' unit and unit H, as the second unit was called, had some contact with the computer but not a great deal.

8.3. ORGANIZATIONAL VALUES

The first research step was to gain an understanding of how different groups saw organizational values within the Inland Revenue. Figures 1 and 2 show some differences of opinion on the organizational values of Centre 1. Both tax officers and systems designers saw the Centre 1 ethos as on the left of the axis with a liking for disciplined employees who would use standardized methods, and placing importance on efficiency and tightly structured jobs. Departmental managers, in contrast, saw the values of Centre 1 as more flexible with an emphasis on the personal qualities of staff. The double mode on task structure indicates that the managers did not agree on whether Centre 1 preferred tightly or loosely structured jobs. However, when the question on the values people *would like* Centre 1 to have, was answered, the profiles of the tax officers and managers were closer to each other than to the systems designers. The system designers thought that organizational values did not require any change: tax officers and managers were split on whether standardized methods should be used or employees encouraged to work out their own methods and use their own discretion.

The systems designers gave a number of reasons for their view that existing organizational values were correct. They pointed out that the constraints of income tax meant that a too loose system of discipline or too many private methods of working were not tenable. The need for equitable treatment of tax payers restricted opportunities for introducing flexible work procedures. They stressed that staff were encouraged to take their own decisions and that the use of individual discretion was encouraged whenever this was possible.

Answers to the question on the "best form of department structure for clerical staff in general" are shown in Figures 3 and 4. The tax officers are in the centre or on the right hand side of the scale on every question, preferring flexible jobs, individual and group discretion over job methods, access to all relevant information and decisions made on the basis of group discussions. There is some divergence of opinion amongst both managers and systems designers on the first question concerning whether jobs should be flexible or clearly defined. Systems designers then move over to the left hand side of the scale on one question only, indicating their

163

Figure 1. Perceived organizational values (modes)

Place a tick where you think this department fits on the scale below.

It has a strong belief in the importance of shared values – that employees should agree with its objectives and the way it carries out its operations

It likes to have disciplined employees who will put the firm's/office's interests first and willingly accept orders and instructions

It prefers to use standardized methods and procedures whenever possible as it believes that this assists efficiency in this kind of firm/office

It places a great deal of emphasis on efficiency and high production and less on personal qualities such as friendliness, trustworthiness, cooperation etc.

It feels the need to organize work activities into tightly structured jobs which are clearly defined and do not permit a great deal of individual discretion

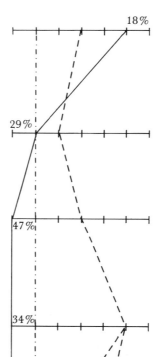

A multiplicity of different values exists and is tolerated here. It is quite acceptable for employees to hold very different views on what the firm should be doing and how it should be doing it

It does not believe in very strict discipline and tries to provide a situation in which people can pursue their own interests

It likes employees to work out their own methods for doing things and to use their own judgement when taking decisions

It places a great deal of emphasis on personal qualities such as friendliness, trustworthiness, cooperation etc. and less on efficiency and high production

It tries to organize work in such a way that employees have loosely defined and structured jobs which permit a great deal of individual discretion

———————— N = 134   Clerks
– – – – – – N = 2   Managers
– · – · – · – N = 3   Systems designers

164

Figure 2. Desired organizational values (modes)

Please tick where you *would like* this department to fit on the scale below.

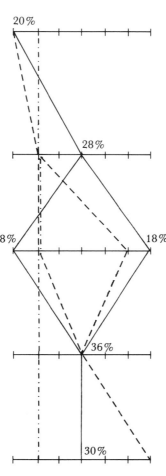

It should have a strong belief in the importance of shared values – that employees should agree with its objectives and the way it carries out its operations

It should try to create disciplined employees who will put the firm's/office's interests first and willingly accept orders and instructions

It should try to use standardized methods and procedures whenever possible as this would assist efficiency in this kind of firm/office

It should place a great deal of emphasis on efficiency and high production and less on personal qualities such as friendliness, trustworthiness, cooperation etc.

It should try to organize work activities into tightly structured jobs which are clearly defined and do not permit a great deal of individual discretion

A multiplicity of different values should exist and be tolerated here. It should be quite acceptable for employees to hold very different views on what the firm should be doing and how it should be doing it

It should have an easy-going kind of discipline and try to provide a situation in which people can pursue their own interests

It should encourage employees to work out their own methods for doing things and use their own judgement when taking decisions

It should place a great deal of emphasis on personal qualities such as friendliness, trustworthiness, cooperation etc. and less on efficiency and high production

It should try to organize in such a way that employees have loosely defined and structured jobs which permit a great deal of individual discretion

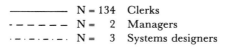

165

Figure 3. Organizational models (modes)

Please tick on the scale below what you believe to be the best form of department structure for clerical staff in general.

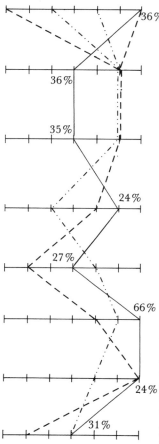

Jobs should be clearly defined, structured and stable

There should be a clear hierarchy of authority with the man at the top carrying ultimate responsibility for all aspects of work

The most important motivators should be financial, e.g., high earnings and cash bonuses

Jobs should be carefully defined by O&M department, management services or supervision and adhered to

Targets should be set by supervision and monitored by supervision

Groups and individuals should be given the specific information they need to do the job but no more

Decisions on what is to be done and how it is to be done should be left entirely to management

There should be close supervision, tight controls and well maintained discipline

Jobs should be flexible and permit group problem solving

There should be a delegation of authority and responsibility to those doing the job regardless of formal title and status

The most important motivators should be non-financial, e.g., work challenge, opportunity for team work

The development of job methods should be left to the group and individual doing the job

Targets should be left to the employee groups to set and monitor

Everyone should have access to *all* information which they regard as relevant to their work

Decisions should be arrived at through group discussions involving all employees

There should be loose supervision, few controls and a reliance on employee self-discipline

```
——————— N = 134  Clerks
- - - - - - N =   2  Managers
-··-··-··- N =   3  Systems designers
```

166

Figure 4. Models of man (modes)

Please tick on the scale below how you see the general characteristics of the majority of staff in this department.

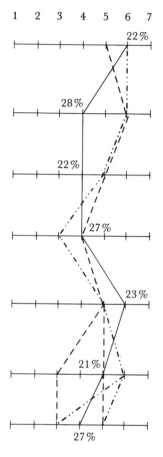

They Work best on simple, routine work that makes few demands of them	They Respond well to varied, challenging work requiring knowledge and skill
They are Not too concerned about having social contact at work	Regard opportunities for social contact at work as important
They Work best if time and quality targets are set for them by supervision	Are able to set their own time and quality targets
Work best if their output and quality standards are clearly monitored by supervision	Could be given complete control over outputs and quality standards
Like to be told what to do next and how to do it	Can organize the sequence of their work and choose the best methods themselves
Do not want to use a great deal of initiative or take decisions	Like, and are competent to use initiative and take decisions
Work best on jobs with a short task cycle	Able to carry out complex jobs which have a long time span between start and finish

```
————————   N = 134   Clerks
· — — — —   N =   2   Managers
— ·· — ·· — ··   N =   3   Systems designers
```

preference for jobs to be carefully defined by O&M, management services or supervision. Managers are on the left hand side of the scale on two factors. They prefer targets to be set and monitored by supervision and they like close supervision, tight controls and well-maintained discipline. It would appear that there are no major value differences between the groups on departmental structure.

A similar closeness of response was found in the answers to the question "how do you see the general characteristics of the majority of staff in this department?". All three groups were either in the centre or to the right of the scale with the exception of the statement on output and quality standards, where the systems designers were slightly further to the left, and the last two statements on initiative and decision taking, and the length of the task cycle, where the managers had different viewpoints.

The systems designers thought that the computer system fitted these flexible staff characteristics reasonably well, primarily because it had been designed to be compatible with the co-existing clerical system. It had removed a great deal of work drudgery by taking over many routine tasks while leaving important areas of decision taking to the staff.

It did, however, impose certain constraints on staff and was sometimes seen by them as too rigid to respond to the very complex problems with which they were working. The managers agreed that the computer removed a great deal of tedious routine work, leaving time for more satisfying work but thought that many staff saw it as restricting their freedom to use initiative and take decisions.

An examination of the answers on organizational values show some interesting differences in perception but few major differences of opinion between the three groups. The systems designers saw Centre 1 as having tight discipline and controls and supported this value position. The managers, in contrast, saw organizational values as more flexible and approved of this. The tax officers saw organizational values as generally a theory Y kind and would have liked them to be more flexible. There was reasonable agreement between the three groups on the best form of departmental structure for clerical staff, and on the characteristics of the majority of staff. A preference was now shown for a flexible structure and staff were viewed as having the talents to fit easily into such a structure.

8.4. WORK VALUES

Managers and systems designers were asked what they saw as most important in producing efficiency in an organization and their answers were categorized according to whether their suggestions were cognitive (related to the use of knowledge), cathectic (the stimulation of emotional responses) or moral (associated with ideas of right and wrong). These are shown in Figure 5.

The emphasis of the answers fits easily into the cognitive, cathectic and moral categories. Efficiency is seen as being created by a rational work environment with accepted objectives, smooth organization of work, good training and a match between what the work required and people's intellectual strengths. Happiness comes as a result of evoking certain emotional responses in staff: letting them feel

Figure 5. Work values (systems designers' and managers' view)

EFFICIENCY

Role expectations
(what is expected of staff)

Cognitive	Cathectic	Moral
	Job satisfaction	

Role requirements
(what the organization must provide)

Cognitive	Cathectic	Moral
Belief in objectives	Creating the right environ-ment	Good communication
Clearly defined procedures		Staff involvement in making decisions
Smooth work flow	Good working relationships	
Work methods allow for the unexpected	Good working environment	
Volume of work matches re-sources		
Good organization		
Clear instructions		
Sound training		
Training in problem solving		
Work content fits people's strengths		

HAPPINESS

Role expectations
(what is expected of staff)

Cognitive	Cathectic	Moral
	Job satisfaction	

Role requirements
(what the organization must provide

Cognitive	Cathectic	Moral
Strengthen personal weak-nesses	Show an interest in people	Consult
Provide integrated tasks	Give recognition	Help career development
Keep crises under control	Let people show what they can do	
	Develop good staff relation-ships	
	Keep staff slightly under pressure	

(Figure 5 continued on page 170)

(Figure 5 continued from page 170)

FAIRNESS AND JUSTICE

Role expectations
(what is expected of staff)

Cognitive	Cathectic	Moral
—	—	—

Role requirements
(what the organization must provide)

Cognitive	Cathectic	Moral
	Show work is being ap- preciated	Senior management is seen as fair and just
		No nepotism
		Management does not ask staff to do what they would not do
		Equal sharing of work load
		Equitable treatment
		Work done by correct staff grade
		Good communication
		Effective grievance machinery
		Balance of power
		Trust
		Honesty

---

that management is interested in them as individuals, will stimulate and recognize their talents and is anxious to develop good relationships. One interesting suggestion was that a slight feeling of pressure is enjoyable. A sense of fairness and justice was associated with various strategies to ensure equity and make staff aware that they were being treated fairly. Staff in the Inland Revenue are under a great deal of pressure because of the amount of work they have to handle and answers reflected this. For example, in regard to efficiency it was suggested that the volume of work should match the available resources; under happiness that crises should be kept under control, and under fairness that the work load should be equitably shared.

Managers and systems designers defined job satisfaction in the following way.

—A person should feel that his talents are being used and that others recognize this fact.

—When a job is finished and you feel that it has been done as well as possible in the prevailing circumstances.

—A satisfying job is one that produces a sense of achievement, one which brings its own rewards (e.g., is of value to the community, full of interest and purpose) quite apart from the financial rewards.

—The feeling that a job is difficult enough to keep someone on their toes without putting them flat on their backs (a systems designer).

170

—At best the officer can't wait to get started, then the day races past, at the end of the day he is reluctant to stop work (a systems designer).

Three of these definitions, two of them from the systems designers, see job satisfaction as an emotional response to challenge.

In order to gain an understanding of the systems designers' values and philosophy in the systems design process, they were asked to rank in order of importance the activities they would include in the phrase "systems design". The answer of each is given below.

Systems designer A
1. Determine the *real* problem
2. Decide whether it is possible to computerize
3. Define input and output
4. Determine broad clerical procedures
5. Determine broad computer processing
6. Define detailed clerical procedures
7. Define computer processing
8. Redefine 6 and 7
9. Evaluate live system from user viewpoint

Systems designer B
1. Determine the objectives
2. Define the system constraints
3. Analyze the task
4. Get agreement on user requirements
5. Design the computer option
6. Devise clerical procedures and instructions
7. Organize training
8. Develop implementation plan

These two lists are similar both in content and rank order.

The systems designers were also asked to describe their design principles and approach. One of them gave a general answer, saying

It is important to ignore what is done now and to look at what needs to be done. Initial emphasis should focus on the user's input and output information requirements; what goes on in the middle can be considered later. The maximum use should be made of machine capabilities.

The second systems designer referred to the Inland Revenue system when he said,

The system had to be designed on a cost effective basis. This meant that to some extent it was designed to suit machine rather than human requirements. If the cost factor had been less important then the system might have been designed to suit the user more; although we did try to give assistance to the user.

It must be remembered that this system was designed at a time when hardware was expensive and clerical labour less so. This situation is now being reversed and would affect costings made today.

8.5. THE DEPARTMENTS WHICH WERE INVESTIGATED IN CENTRE 1

Four departments were chosen for investigation, two of them using the computer and two still largely operating manually. The two computer-based departments were allocation units 3 and 4; the non-computerized departments were unit H which looked after tax payers with complex financial affairs and the employers' unit which kept employers' records and liaised with employers.

Centre 1 handles the affairs of P.A.Y.E. tax payers. Prior to the introduction of the computer the responsibility of an allocation unit was to assess income tax under schedule E, to issue and receive tax returns, to calculate P.A.Y.E. codings and schedule E assessments, to keep a record of where the tax payer was employed and to process the employers' tax deductions. An essential part of the job was determining whether income was assessable, whether allowances were due and the amounts of both. Tax officers kept in direct contact with tax payers and employers by means of letters, phone calls and personal visits.

The introduction of the computer did not make a great deal of difference to the work of the allocation unit. The tax officer continued to have an allocation of tax payers' files to look after, to correspond with tax payers and to determine what allowances and reliefs were due to them and what deductions were to be made. The difference was that having made a decision he then notified the computer which amended the tax payers' record and issued the amended notices of coding to the tax payer and his employer. The computer removed a great deal of routine work from the tax officer, particularly that associated with a tax payer changing jobs and the need to transfer his file from one employer to another. This linking was now done automatically by the computer. One problem which stemmed not from the computer, but from the decision to build a large centralized office was the removal of the tax officer from face-to-face contact with the public. Many members of the public had welcomed the opportunity of visiting their local offices and talking personally with their tax inspector when they were experiencing tax problems. The location of all tax offices and records for Scotland in East Kilbride meant that it was now extremely difficult, or even impossible, for members of the public to visit the Centre. This interaction was missed by both the public and the tax officers.

The employers' unit, although not greatly affected by the computer in its work, had been set up as a consequence of the change-over to computerization. When tax payers' records had been identified according to the employer they worked for, the tax officer had close day-to-day contacts with a small group of employers. The change-over to a computer and to identifying the tax payer by his national insurance number, not his employer, removed the need for this contact and there was therefore the necessity to create a special unit which would liaise with employers. The employers' unit now held the employers' files and its staff were responsible for disseminating to the allocation units all the information relevant to individual employees. Staff in the employers' unit also dealt with all forms completed when a tax payer changed employers and it was their responsibility to refer

172

these forms to the appropriate files in the allocation units. They also assisted with any problems employers might have.

Unit H dealt with complex tax cases and because its work involved difficult assessments and negotiations it employed a number of case workers. Figure 6 shows the distribution of tasks between the tax officer and the computer.

8.6. GOAL SETTING AND ATTAINMENT

The principal reasons for introducing the computer were difficulties in recruiting high calibre staff, the fact that taxation work was increasing and becoming more complex and the availability of a technology that could take over a great deal of the routine work and at the same time secure some savings of staff. The systems designers associated with the project in its early stages saw the principal technical objective as computerizing the manual system without in any way changing its philosophy. Because the manual system record cards held information for a period of six years, the computer too must be able to produce a historical record document. Decisions on what to computerize and what to leave for the tax officers to carry out were greatly influenced by the hardware available at the time. This could only handle so much information during the day and so a decision was taken not to put complex cases on the machine.

The main business objective was to improve the service to the tax payer. It was believed that faster processing plus fewer errors would produce a better service. Human objectives were of two kinds. First, there was the need to save staff and it was necessary to do this in order to pay for the computer which required £330,000 capital outlay. A second objective was to improve the quality of work for staff by removing routine. This objective was made clear in a booklet, "Looking ahead to East Kilbride", produced by the automation liaison committee in 1967 as a guide to staff who would be joining Centre 1. Job satisfaction as a system objective was encouraged by the interest of the Inland Revenue Staff Federation in obtaining the best work and working conditions for their members. Centre 1 provided an opportunity for doing this.

Despite their previous experience with the Inland Revenue, many of them as tax inspectors, it was not an easy system for the systems designers to design. An inherent conflict arose from the fact that the system must be designed in the best way for the computer and also in the best way for the man at the desk and these two objectives did not always coincide. For example, the removal of the tax payers' history from the tax inspector to the computer made him feel that he had lost information. Systems designers also experienced the constraint that is normal in civil service applications which affect the public: the need to give tax payers whose records were on the computer exactly the same service as those who were still dealt with manually. This meant that tax payers in Scotland, who, once Centre 1 was established, would all be on the computer, had to be dealt with in the same way as all other tax payers. This produced a number of design difficulties. Items such as expenses which were easy to deal with manually were initially less easy to handle on the computer, and the coding of jobs for computer records caus-

Figure 6. Outline of daily processes

Process	1	2	3	4	5	6	7	8	9	
Tax payer Write letter			X							Inputs to
Complete return						X				Centre 1
Employer Send DC								X		
Send P45 or P46									X	
AO Deal with letter			A							Action decision
Examine return						A				input to computer
Prepare input form			F			F				
Computer updated record			F			F		A	A	(AO can do all
Compute coding	O			F					F	these manually)
Compute assessment		O				O		A		
Compute repayment		O				R		O		Computer
Issue P2 & P6	O			O				O	O	processing
Issue P70		O				O		O		and
Issue return									O	computer output
Issue repayment		R				R		R		
Issue printout		O		O		O		O	O	
AO File copy P2	O			O				O	O	
File copy P70		O				O		O		
Check assessment		R				R		O		Action decisions
File repayment advice		R				R		R		(usually by comple-
Release repayment		R				R		R		ting new input form)
Deal with printout		O		O		O		O	O	
Write to tax payer	R	O	R	O	O	O	O	R	R	
Employer										
Operate P6 code	O			O				O	O	
										Outputs from
Tax payer receives P2	O			O				O	O	Centre 1
P70		O				O		O		
return									O	
repayment		R				R		R		
letter	R	O	R	O	O	O	O	R	R	

Initial action = X; Consequential action: A = always, F = frequently, 0 = occasionally, R = rarely.
AO = Allocation officer (tax officer); P2 = Notice of coding; P6 = New code for employer; P70 = Assessment; DC = Deduction card.

174

ed problems. There were also legal constraints. For example, it is a legal requirement that tax assessments are made by a tax inspector and tax returns must be sent to the tax payer at his home address. Also, names had to be printed on the tax returns and sticky labels were not permitted; employers must receive paper forms giving the code numbers of their employees. A constraint of a different kind was the fact that the Centre 1 building was being designed at the same time as the computer system and the two had to fit together and be ready simultaneously for the new staff transferred from other parts of Scotland.

The first report on computerization was prepared by a specially set up "committee on centralised automatic data processing". Its Chairman was the Deputy Chief Inspector for P.A.Y.E., Mr. Barford, and it was therefore known as the Barford Report. This was a high-powered committee containing both Board and user representatives and its task was to examine the possibility of using a computer to centralize P.A.Y.E. records, and to advise on the stepts by which such centralization might best be achieved. The committee made a careful investigation of available hardware, costs and the design of computerized records, and visited the USA to look at similar systems there. They produced their report at the end of 1962. The Inland Revenue Staff Federation and the Association of Inspectors of Taxes were interested in the Report and soon after its release a management-union liaison committee was set up. Finally a design team was selected. This too visited the United States to find out the kinds of technical difficulties the Americans had experienced. An outline for a possible system was then prepared.

8.6.1. Design Alternatives

The design of the Inland Revenue computer system was based on an evolutionary approach. The systems designers had a reasonably free hand, although they had to sell their ideas to the user, and their decisions were influential during the setting up of the ADP system and the transfer of records from the manual system. This change-over took two years and was accomplished through a gradual run down of district offices by transferring major employers to the computer first. Because of the expense and limitations of the hardware available at the time, there was a pressure to fit the system to the capabilities of the computer rather than the needs of the tax officers. The systems designers felt that they must play safe and could not afford to run the risk of part of the system not working for technical reasons. Today, with more flexible and technically sophisticated hardware available, a more user-oriented approach could be adopted. In retrospect the systems designer who was associated with the initial design of the system said that he would have preferred to create a simpler system, but to do this would have required a simplification of the tax structure.

Despite the constraint imposed by the technical limitations of the hardware, a great deal of thought was given to the needs of the tax officers. One important problem that had to be considered arose from the fact that the computer would remove work from the tax officers; they would therefore need to deal with more tax payers and this would require some reorganization of their work. In the manual system the tax officer had dealt with groups of employers, but this had meant that whenever a tax payer changed his job his file had to be physically moved from the tax officer dealing with his previous employer to the tax officer dealing with his new employer. This movement of files caused problems with files becoming lost or delayed in transit. It would have been cumbersome to transfer this system to the computer and the decision to identify tax payers by their national insurance number, and not their employer meant that a tax officer could deal permanently with the affairs of a group of tax payers, providing they did not move from Scotland. This change meant that tax officers who had become very familiar with groups of employers now lost this relationship and had to deal with tax payers employed in a wide range of different jobs, some with very complex tax situations. This increased the complexity of Revenue work and presented training problems. The method for reducing this complexity eventually adopted was the transfer of particularly difficult cases to a special unit, unit H, which was staffed principally by higher grade tax officers. The movement of tax payers from one job to another was dealt with by the specially created employers' unit. The function of this unit was to ensure that an employer was operating P.A.Y.E. correctly and giving his employees the information they required.

The systems design team did see it as part of their job to consider job satisfaction and to create interesting jobs for tax officers and they made every effort to do this. The impact of the computer on job satisfaction was discussed in detail in the document "Looking ahead to East Kilbride", produced by the automation liaison committee in 1967. The concluding paragraph of this document reads:

> The growing use of computers is an essential feature of our developing civilisation; it is a move towards the liberation of the human mind from mechanical tasks which, for want of adequate machines, it has had to perform for so long. In the scientific and technical field computers open the way for ambitious and far-reaching projects based upon calculations and control processes not previously within human capability. In commerce they offer release from time-consuming calculations and drudgery and open possibilities of concentrating upon more rewarding activities. The Inland Revenue is now part of this general movement, and in the sheer size and scope of its projects, a leader. The Department has a loyal and efficient staff which for many years has struggled with an ever-growing volume of clerical work. Now at last there is the prospect that much of this volume can be transferred to machines, leaving the staff better able to use their skill and energy in ways more befitting intelligent human beings.

The realization of this clear statement of values was achieved by removing routine work from the tax officer and leaving him with problem solving and interaction with the tax payer. In addition units were organized into teams so that a small group of tax officers would work together on the problems of a group of tax payers. Further humanization of work was restricted by the technology. A major aid to the tax officers' job would have been easy and immediate access to the tax payers' record. This required an on-line system and VDUs which were not available at the time. These are being reconsidered today in conjunction with a decentralized form of organization based on district tax offices. The systems design group's interest in job design was reinforced by the Inland Revenue Staff Federation's desire to ensure that its members had challenging jobs and good working conditions. The staff side were able to make their views known through their representation on the computer liaison committee and their contributions were very helpful to the systems designers.

The design alternatives that could be considered in the Inland Revenue were inhibited by the existence of the manual system and the need for the computer system to fit tightly with this so that a uniform tax service could be provided throughout the country. But it was the only one of the government departments investigated in this research to make job satisfaction and the creation of interesting work a design objective and to make clear the values which lay behind this approach.

8.7. ADAPTATION

The difficult process of first blending an existing manual system with a new computer system and then gradually replacing manual procedures with computer procedures was assisted by the creation of a number of committees. A liaison committee was formed in 1963 which brought together the systems designers, the users, their staff representatives and the Board of the Inland Revenue. This committee continued until 1968 and set up a number of sub-committees. The knowledge of the committee members, in particular of the two staff side members, proved extremely useful to the design group. The staff side were able to put forward the view of the tax officers and the systems designers' own experience as tax inspectors helped them to understand this.

A steering committee was set up with responsibility for ADP in the Department of Inland Revenue, and this still exists today. A number of user committees were also established to consider detailed aspects of systems design such as procedures and documents. Once Centre 1 was staffed then a comprehensive structure of Whitley committees became available to assist the adaptation process. Today, in addition to the main committee, Whitley committees cover management procedures, accommodation and hours and welfare. The Whitley structure covers staff grades up to executive officer level, while the Inland Revenue Staff Federation covers most grades and concentrates on national issues.

During 1965 and 1966 a series of twelve articles appeared in the Inland Revenue Federation journal, *TAXES*. These articles discussed the functioning of the new Centre 1 and had as their objective keeping staff informed of developments at East Kilbride, and giving them an understanding of ADP and computers. The plan was to start the Centre with a small advance group of staff who would test the new system out on a comparatively small number of files, around 30,000, prior to the start of the main build-up of staff and files. Files and staff would then be brought in gradually over two years until the full Centre complement of 1,300 staff had arrived.

A comprehensive training programme was developed and designed to operate in three phases. Phase 1 covered a one day general appreciation course directed at introducing tax officers to the subject of computers, the organization of Centre 1 and the processing of P.A.Y.E. by computer. A mobile training unit toured Scotland between April 1965 and April 1966 so that all staff who would be transferred to Centre 1 could participate in this introductory course. Phase 2 was training for the advance party. During the first nine months of the setting up of Centre 1 the advance party of 180 staff were sent on training courses to give them the skills to transfer the records of employees and employers to magnetic tape and generally to operate the new system. Phase 3 training was for the main complement of staff and immediately before their arrival at Centre 1 they were sent to one of the Inland Revenue's training centres. Once the Centre was established new staff were inducted with the help of the Centre's own training unit. Experienced tax officers were given courses of about two weeks; the training for new entrants to the Inland Revenue service lasted around twelve months.

Because many staff were moving to East Kilbride from other parts of Scotland the East Kilbride Development Corporation made accommodation of different kinds available to Inland Revenue staff for rental. East Kilbride itself was an attractive new town within commuting distance of Glasgow and with excellent shopping and sports facilities.

8.8. INTEGRATION

Once staff had arrived and been trained the new Centre had to be welded together into an efficiently functioning unit which gave job satisfaction to its staff. This integration was not easy to achieve. First of all, staff who had been used to working in smaller offices in other parts of Scotland had to be brought together and provided with a situation in which they could work cooperatively. Second, in addition to this complex human integration problem, the human part of the system had to be integrated with a new technical system. Third, the integration of tax officers with their tax paying clients might prove a problem, for the centralization of records in Centre 1 meant that it would be difficult for tax payers to make personal visits and communication would therefore be restricted to letter or telephone.

8.8.1. The Impact of the Computer on the Structure of Work

Computer systems, if they have not been designed to achieve human objectives related to job satisfaction, can adversely affect jobs through making them more routine, or more tightly controlled and with fewer opportunities for the use of discretion. But the Inland Revenue systems designers had tried hard to cater for the job satisfaction needs of staff through the way they designed the computer system and allocated tasks between the machine and the tax officers. They made it their objective to provide the following:

—interesting and varied work;
—opportunities for personal growth through acquiring new skills and knowledge;
—opportunities for taking decisions;
—opportunities to undertake reasonably large jobs;
—opportunities for solving complex problems;
—opportunities to develop new methods of work on own initiative;
—freedom to set own time and quality targets.

They pointed out, however, that tax officers have always had these opportunities and that by removing routine work they may have increased the pressure under which the tax officer worked. Tax officers at Centre 1 had to deal with three times the number of tax payers than a manual centre would have handled, and because routine had been removed and problem solving left, the complexity of work had increased. The provision of face-to-face contact between the tax officers and the public had also been removed. Figures 7 and 8 show the work flow and work responsibilities of one of the ADP units.

The job descriptions in Figures 7 and 8 and the job structure comparisons in Table 1 show that work has not been routinized by the computer. Both tax officers – higher grade and tax officers in the allocation units – have jobs which meet our criteria of good job design. The clerical assistant group whose work is routine, is the result of an organizational decision in unit 3 to create a service group with responsibility for activities such as filing. The routine nature of the work is nothing to do with the computer.

8.8.2. The Clerks' Perception of Integration

The tax officers in the four units completed the job satisfaction questionnaire to establish what they liked most and least about their work and how they responded to the impact of the computer. The author received a great deal of help from management and staff representatives in carrying out this survey. Although completion of the questionnaire was entirely voluntary, staff were encouraged to complete it and the author was introduced to the tax officers in each unit by the Staff Federation representative. Despite this assistance there is likely to be some bias in the results from the employers' unit as only 9 members of staff completed the questionnaire. The survey population in the employers' unit and unit H was 30, and one half of these were thirty years of age or less.

51 staff in unit 3 and 52 staff in unit 4 completed the questionnaire. 72 of these were members of live working groups and the rest were responsible for movements or ancillary activities. 78% were thirty years of age or less (Tables 2 and 3).

The data show few differences between the manual and the ADP units and those differences that do appear show the staff in the allocation units to be more satisfied with certain facets of their work than staff in the employers' unit and unit H. For example, allocation unit staff were less likely to say that there was too

Figure 7. Workflow ADP unit
(This unit deals with 420,000 tax payers' files)

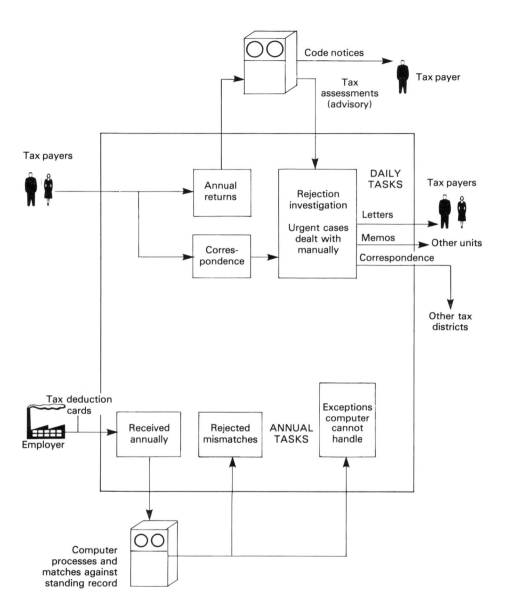

Figure 8. Typical ADP unit live working group*
(a group which deals with tax payers)

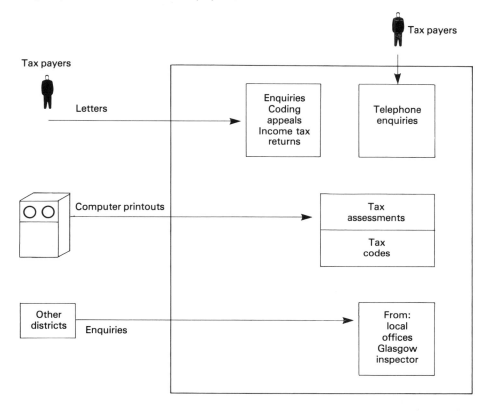

* The group consists of one tax officer higher grade and 6 tax officers. It is responssible for a band of 10 national insurance numbers and has to ensure that these tax payers pay the correct tax, and have the correct tax codes.

Table 1. The government department. Job structure comparisons

CONTROL LEVELS

Definition of skills:
A. Communicating in writing
B. Communicating verbally
C. Arithmetical skills
D. Machine operating skills
E. Checking/monitoring/correcting
F. Problem solving
G. Coordinating
H. Supervising

	OPERATIONAL LEVEL												HIGH CONTROL LEVELS					
	Reduce boredom	Variety — Give sense of personal control	Means choice — Use of judgement / Knowledge choice — Use of initiative — Give feeling of achievement	Job identity — Give sense of making important contribution	Job identity — Give sense of team work	Job relationships — Give sense of confidence	Goal clarity — Problem prevention	LEVEL 2 Anti-oscillation — Efficiency improvement	LEVEL 3 Optimisation — Creativity	LEVEL 4 Development — Autonomy	LEVEL 5 Overall control — Key task
Employers / Unit	No. of tasks / No. of skills	of methods	of work sequence / Use of initiative	Clearly defined start to job / Clearly defined end to job / Uninterrupted task sequence / Long task cycle 20+ minutes / Visible contribution to product or service	Works as member of group	Considerable inter-action required / Clear work objectives	Objectives not too easy or too difficult / Can requisition required resources	Can correct errors, solve problems / Coordinates own work activities	Coordinates group work activities / Can improve methods	Can improve product or service / Individual free from supervisory control	Group free from supervisory control / Problem solving
Unit H. Employers Unit. manual											
Tax officer higher grade – case worker	Large 4 A/B E/F	✓	✓ ✓	✓ ✓ ✓ ✓ ✓	✓	✓ ✓				Problem solving	Problem solving
Allocation units 3 and 4. ADP											
Tax officer higher grade – supervisor	L 6 A/B/E F/G/H	✓	✓ ✓	✓ ✓ (with some cases) ✓ ✓	✓	✓ ✓	✓	✓ ✓	✓ ✓		Problem solving / Organizing / Supervising
Tax officer	L 5 A/B/E F/G	✓	✓ ✓	✓ (for part of job) (with some cases) ✓ ✓	✓	✓ ✓	✓ ✓	✓	✓		Problem solving / Data input and output
Clerical asst. (in service group)	S 1 A		✓	✓ ✓	✓	✓					Collecting information

Note: ✓ in the "These are the day to day operating tasks" column is marked for each job row.

Table 2. Government department. Manual and ADP unit opinion on job structure (in percentages)

Category	Measure	Type	Item	Manual	ADP 3	ADP 4
VARIETY	Routine and challenge	FACT	Work is usually or always routine	40	36	25
VARIETY	Routine and challenge	PREF	There is too little challenge	20	12	20
VARIETY	Routine and challenge	FACT	There is too much routine	75	45	55
VARIETY	Routine and challenge	PREF	Almost everything has rules or set procedures	63	40	55
DISCRETION	Means choice	FACT	Can hardly ever choose our work methods	20	4	17
DISCRETION	Means choice	FACT	Can hardly ever choose work sequence	13	10	16
DISCRETION	Means choice	FACT	Can hardly ever choose next task	31	11	22
DISCRETION	Means choice	PREF	Would like more choice of work methods	50	28	34
DISCRETION	Means choice	PREF	Would like more choice of work sequence	44	26	50
DISCRETION	Means choice	PREF	Would like more choice of next task	48	29	38
DISCRETION	Knowledge choice	FACT	Can hardly ever make decisions, use judgement	13	4	12
DISCRETION	Knowledge choice	PREF	Would like more opportunities to make decisions, use judgement	63	39	47
JOB IDENTITY	Task cycle and integration	FACT	Has very little or little chance to see a piece of work through from start to finish	3	14	10
JOB IDENTITY	Task cycle and integration	PREF	Would like more chance to do this	79	98	91
JOB RELATIONS	Work dependency	FACT	Inefficient work by others can almost always or usually hinder me	41	44	30
JOB RELATIONS	Work dependency	FACT	Inefficient work by me can almost always or usually hinder others	36	44	45
JOB RELATIONS	Work dependency	PREF	I would like there to be less work dependency	37	24	38
GOALS	Deadlines	FACT	There is too little or barely enough time to do work	73	64	72

Table 3. Government department. Positive and negative feelings about work
(in percentages)

Positive feelings about work

	PREF	FACT	FACT	FACT	FACT	FACT	FACT
		Time never seems to drag	My present job is close to my ideal job 5/6	I am strongly involved in my job	Taking job as a whole I enjoy it very much	I would like to get sense of achievement more often	I get sense of achievement almost every day
Manual		30	20	–	20	93	17
ADP 3		42	15	14	35	78	45
ADP 4		21	26	17	31	91	33

Negative feelings about work

	FACT	FACT	FACT	FACT	FACT
		Time drags ⅔ of the day or more	My present job is far from my ideal job 1/2	I am very little involved in my job	Taking the job as a whole I do not enjoy it	I hardly ever get a sense of achievement
Manual		14	33	8	20	30
ADP 3		6	17	4	4	16
ADP 4		16	28	10	8	18

184

much routine and fewer of them wanted more opportunities to take decisions and use judgement.

Staff in the employers' unit and unit H were less likely to say that they got a sense of achievement from their work every day and that they were strongly involved in their jobs. They were also more likely to say that "taking their jobs as a whole, they did not enjoy them". These data provide no evidence of the computer having had any disadvantageous impact on the work of the tax officers in the allocation units.

In order to gain an understanding of what the tax officers liked about their work they were asked what kinds of decisions they most enjoyed taking. The table below shows that a considerable number of clerks in each group wanted the opportunity to take more decisions and use more judgement.

Answers fell into the following categories:

What kinds of decisions do you most enjoy taking?

Planning my own work	Deciding how to handle cases	Casework decisions	Management decisions	Other
12	9	31	11	7

Tax officers spoke of liking complex cases, particularly those which required some interpretation of the tax law (unit H), and of valuing opportunities to use judgement and take responsibility.

They were asked what gave them the most sense of achievement and their answers fell into the categories shown below:

What gives you most sense of achievement in work?

Finalizing complex cases	Helping the public	Working efficiently	Keeping up to date	Other
34	28	16	23	3

Many staff referred to the feeling of achievement they received from solving difficult problems and cases satisfactorily and 28 referred specifically to their pleasure at being able to serve the public.

—Being able to give the tax payer good service.
—Doing a job well and keeping the tax payer satisfied.
—Trying to keep the public happy and being of some service to them.
—Feeling a tax payer appreciates your work.

Most of the other answers could be categorized as working efficiently or keeping up to date with the work.

Answers to the question "what causes time to drag most of all?" referred almost entirely to routine work. Although it can be argued that a certain amount of routine reduces the pressure of work, many tax officers would have liked less.

Tax officers were asked how their work could be made (*a*) more efficient and (*b*) more enjoyable or satisfying. A very large number of replies were received to both questions. The points made most frequently in answer to the efficiency question concerned the need for a lighter work load and more staff (17 suggestions). Also, better use of the computer, either through an on-line facility providing immediate access to records, a simplification of computer programmes and printouts or the transfer of more routine work to the computer (15 suggestions); and fewer interruptions, particularly telephone calls from members of the public (14 comments). All these suggestions were related to either reducing or streamlining work. The response given most frequently to the question on enjoyment was similar for the most important contribution to enjoyment and satisfaction was seen as a smaller work load (16 comments). Tax officers also said enjoyment would be increased through larger jobs, more discretion and less routine work (11 suggestions) and through better working conditions (9 comments).

Tax officers were also asked whether the use of a computer by their department increased or decreased their efficiency and the interest of their work. The answers are shown in Table 4.

Table 4. Do you find that the use of a computer by this department:

	Increases your efficiency		Makes no difference		Decreases your efficiency		Total	
	N	%	N	%	N	%	N	%
Unit H, Employers	5	18	16	57	7	25	28	100
Unit 3	30	60	17	34	3	6	50	100
Unit 4	29	58	13	26	8	16	50	100
All	64	50	46	36	18	14	128	100

	Increases the interest of your work		Makes no difference		Decreases its interest		Total	
	N	%	N	%	N	%	N	%
Unit H, Employers	2	7	25	89	1	4	28	100
Unit 3	16	33	29	60	3	6	48	100
Unit 4	14	29	28	58	6	12	48	100
All	32	26	82	67	10	7	124	100

In unit H and the employers' unit, where the impact of the computer was slight, a majority of staff saw the computer as making no difference to their efficiency. In units 3 and 4 the majority view was that it increased efficiency and tax officers gave a variety of reasons for this result. For example,

—It provides instant information as to whether the file I am looking for has been transferred out or not.
—By relying on the computer to clear simple cases, which would otherwise require my time, I am able to concentrate on important matters which require thought.
—By taking away the long laborious jobs such as coding and assessing, it gives one time to concentrate on the more difficult cases.

But there were some criticisms:

—Without the computer my experience in the practical side of the job would be greater. The presence of the computer tends to create mental laziness.

A majority of tax officers saw the computer as making no difference to the interest or enjoyment of their work. Those that said it increased the interest of work gave the following reasons:

—It increases my interest in my work since the computer deals with the bulk of the routine cases and leaves a greater part of the day for more involved work which holds my attention.
—It keeps you alert as not only do you have to be knowledgeable of income tax but also of the capabilities of the computer.

The Inland Revenue computer system was therefore an example of a technology that was well integrated with the needs of the people who used it. It freed them from a great deal of routine work and enabled them to spend more of their time dealing with interesting and complicated tax cases which the computer did not handle. Internal socio-technical integration in Centre 1 was therefore good but there were difficulties in achieving a similar level of integration between Centre 1 and the tax payer environment. Tax officers had a great deal of work and this subjected them to some stress. They were expected to handle the affairs of far more tax payers than was the case with the manual system. Also, the decision to centralize the tax services for Scotland in Centre 1 had removed opportunities for face-to-face contact with the public which both tax officers and the public had appreciated and enjoyed. All of these problems will be resolved in the future. It is now Inland Revenue policy to retain the existing decentralized system of local tax offices, and technical developments will enable each of these offices to have its own computer link with interrogation facilities provided via visual display units.

8.8.3. Management's Perception of Integration

8.8.3.1. *The Fit with the Needs of the Tax Officers*

The system designers saw the main strengths of the tax officers as their breadth of knowledge, drive, initiative and ability to withstand pressure, but they disagreed on whether the computer system had been designed to exploit these strengths. One thought that it had, the other that the legal constraints and need for uniformity throughout the country had imposed restrictions on the extent to which the system could be designed to meet user needs. They both thought that the system had been fairly successful in meeting the job satisfaction needs of staff. It had removed some work drudgery but it had also made the job more demanding by giving the tax officer more, and more complex, cases to handle. They believed that tax officers valued the improved access to information and the fact that budget alterations could be made quickly and easily by the computer. They suggested that younger staff were more in favour of computer technology than older staff and this was supported in the survey results for younger staff in unit 3 were more likely to say that the computer increased their efficiency than older staff. One problem was that few people really understood the system and how it worked, and to remedy this tax officers were encouraged to take their queries to the Centre 1 computer unit. The main complaint from users was a need for quicker access to information. With the manual system the tax officer had a summary of a tax payer's record in his desk. He or she now had to request this information and the current batch system took a minimum of 24 hours to produce it. This delay would be removed once an on-line system was introduced.

The managers of the four units saw their staff as of high calibre, with high morale, and able to handle a heavy work load and accept responsibility. They saw the computer as making a major contribution to staff needs in a number of different work areas. It carried out the recoding of tax payers which had been an enormous job in the past, it also worked out assessments and sent out coding notices. All of these activities saved staff a great deal of time and gave the public a better service. They felt that their staff missed the contact with the public, although this was nothing to do with ADP. Another problem was the vast amount of paper produced by the computer which tended to create rather than save work. The managers said that their staff had always had interesting and varied work and the computer had, in theory, increased the opportunity for this by providing more time for correspondence and complex cases. Unfortunately high labour turnover at Centre 1 meant that a great deal of time of experienced staff was taken up with teaching and supervising trainees.

Both unit and senior managers said that staff still had problems in understanding the computer system. Most of the staff in Centre 1 had no experience of the manual system and therefore nothing to unlearn, but many of them were inexperienced with 40% in their first year as a tax officer.

188

8.8.3.2. *The Computer's Contribution to Better Management Information and Control*

The systems designer's view was that top management had gained important information as a result of the computer system. For example, the computer could tell them how many cases had not been dealt with the previous year and provide other macro-statistical information. Departmental managers thought that they had lost some information, in particular the immediate access to historical information provided by the manual system record card. They said that the computer had led to more centralized control of the Inland Revenue but that in compensation they now had more accurate information and information that was processed more rapidly. This would be still further improved once VDUs were introduced.

Senior management confirmed that they now had more information. This helped them to achieve better internal control through their ability to compare the performance of the various units and to feed back information on trends to the units.

8.9. MANAGEMENT'S EVALUATION OF THE SYSTEM

Systems designers and managers were asked what aspects of the Inland Revenue computer system best measured up to their idea of a successful system. The systems designers replied that it had been a relatively inexpensive system and that since its introduction it had been developed to take over an increasing number of routine jobs. For example, the computer would now check and correct errors such as incorrect national insurance numbers. They said that, given the existing complex tax structure, they would use the same design approach today. They thought some improvements could still be achieved in the movement of files when employees changed their jobs and that an interactive system using terminals would provide many benefits. A simplification of the tax system would greatly assist the more efficient use of the computer but this would require a top level government decision. In their opinion the change in policy away from large centralized offices to smaller decentralized offices was a move in the right direction. Management also thought that it was a good system but said they would like to have direct access to the tax payer's record instead of a 24-hours wait for information.

The view of the systems designers was that the technical objectives that had been set at the start of the design process had been reasonably well-attained. The computer processed records on a daily cycle, a time scale insisted on by the Treasury, although this had been a struggle to achieve in the early days.

Systems designers and managers were agreed that the early business objectives had been achieved. A uniform service was provided to tax payers and the majority now received a better service although the more unusual or erratic tax payer still had to be handled manually and was not helped by the computer. Staff saving objectives had also been achieved and staff requirements had been reduced by 600 without any damage to the level of tax payer service. They said that if they were designing the system today they would provide direct access to the computer record and make the switch over from one system to another more gradual. There was a need to convince staff that the computer would help them and they had to be given confidence in new technical developments.

8.9.1. The Computer System and Corporate Strategy

Senior management thought that the computer had assisted business strategies in the Inland Revenue by providing management with useful information and by achieving staff savings. They were now able to plan more successfully because computer-imposed deadlines forced management into long-term planning and imposed a discipline; also, the information provided by the computer assisted management in forecasting the future. In addition, the macro-information derived from the computer enabled the Inland Revenue administration to advise the government on tax policy. They pointed out that it was not easy to distinguish the influence of computerization on strategy from the influence of centralization.

8.10. CONCLUSIONS ON THE INLAND REVENUE AND CENTRE 1

8.10.1. Goal Setting and Goal Attainment

Technical, business and positive human goals related to the maintenance and improvement of job satisfaction were set at the start of the design process. All of these were successfully achieved. Business goals were directed at introducing a computer system which provided a stable and uniform service to tax payers throughout the country, and at saving staff. Staff saving was required for two reasons: first, to meet the Treasury's cost-benefit requirements for new computer systems and, second, to reduce the need for staff at a time of labour shortage. Human goals were seen as extremely important, for tax officers have always had interesting and challenging jobs and the computer must not jeopardize the attainment of job satisfaction. If possible it should increase this through removing routine work from the tax officer. This was also achieved. There were few complaints about the computer. A majority of staff thought that it assisted their efficiency and that it had not affected the interest and enjoyment of their work. Goal setting and attainment were therefore very successful.

8.10.2. Adaptation

Adapting staff to the new computer system was made more difficult by the fact that they also had to adjust to a new centralized tax office, Centre 1. They therefore had to assimilate a new form of organization and a new relationship with the public, as well as a new technology. Adaptation to the new technology was assisted by an appreciation course before staff left their previous jobs and training courses immediately prior to their arrival at Centre 1. Adaptation to a centralized organization was helped by the provision of a new, specially built, office complex in an attractive new town, East Kilbride, and by a comprehensive structure of

Whitley committees. The changed relationship with the public was more difficult to handle and face-to-face contacts had to be replaced with letters and telephone calls.

8.10.3. Integration

The integration of the new computer system into the work processes of the Inland Revenue was achieved without much difficulty and the technical and human systems complemented each other. Adjustment to the new, large, centralized tax centre proved more difficult and labour turnover was high. The integration of the tax officers with the public was not satisfactorily achieved from the tax officers' point of view. Because the public could not visit East Kilbride in person, many of them telephoned and these telephone calls were a constant source of interruption. They also missed the opportunity of giving advice personally to members of the public who had previously visited their local tax office whenever they had problems.

8.10.4. Pattern Maintenance

The Administration of the Inland Revenue recognized these problems of organizational and environmental integration at an early stage and a decision was taken not to continue with the concept of large centralized centres throughout the country. There is therefore no intention of extending this particular pattern of operation. The new pattern will seek to link staff in local offices to regional computers through visual display units on their desks. If Ministerial approval is given for the Inland Revenue's present plan, an interactive, on-line computer system will reinforce the already good socio-technical integration by providing tax officers with immediate access to tax payers' records.

9. The International Bank

In this situation, as in the two industrial firms, the researcher was able to make a before and after comparison in the user department.

The author's introduction to the international bank took place in 1974. She was then approached by a senior member of the bank's personnel department who told her that the bank's foreign exchange department was about to move from a batch to an on-line computer system. Because the bank had been experiencing some difficulties in attracting and retaining staff there was a desire to associate this technical change with a reorganization of work which would increase the job satisfaction of staff. The bank therefore proposed to make an increase in job satisfaction a specific system objective and to give the bank clerks themselves responsibility for designing an optimal form of work organization into which the computer system could be fitted. The bank had already created a "job satisfaction working party" from a representative group of foreign exchange clerks and asked the author if she would assist this group to identify their needs and develop a work structure that would meet these needs. The author was delighted to accept this invitation.

The exchange department in the international bank forms part of the exchange and money market division which consists of the chief manager's establishment, the dealing room and the exchange department. The technical activities are performed in the dealing room while the exchange department is an administrative unit mainly occupied with the processing of deals entered into by the dealing room. The majority of the transactions consist of buying and selling foreign currency by means of telephone communication to and from agents all over the world. Currency is also accepted on deposit from other banks and institutions. Dealing is organized by currency rather than transactions and the dealers are split up as follows: dollar section, swiss franc section, deutschemark section, continental section for other currencies, and the sterling section.

The dealing room is staffed by dealers and position clerks. Position clerks keep a record of funds going in or out (the position) and know the value date of deals. Most position clerks will eventually become dealers themselves.

The organizational design project was to be primarily concerned with the exchange department as this had many more routine activities than the dealing room. Whereas dealers were essentially entrepreneurs who used considerable flair and skill in the buying and selling of currency, the work of the exchange department was controlling the paperwork, although there was an element of customer contact. Management believed that the existing functional organization of work was underutilizing the talents of many clerks. Exchange department staff consisted of 68 clerks of whom 25 were part-time. It had a tall grading structure ranging from grade 1 typists to grade 7 assistant managers.

9.1. THE WORK OF A FOREIGN EXCHANGE DEPARTMENT

Dealing staff are responsible for buying and selling foreign currencies and for the application of rates of exchange to all transactions involving currency operations. The dealers are responsible for the department's currency "positions" and for the maintenance and distribution of the bank's currency balances with its overseas correspondents.

In any free market, supply and demand are the principal factors in determining price. These factors also play a major part in arriving at the current exchange value of any currency. If there is a demand for U.S. dollars against sterling, then, other things being equal, the tendency will be for the dollar rate to harden and in exchange for £1 a buyer will get fewer U.S. dollars. Exchange rates respond very quickly to the effect of any disequilibrium in supply and demand and also to news of political or economic importance, whether at home or abroad.

Although bankers refer to the London Foreign Exchange Market, there is no central market as such. The "market" consists of dealers, operating in the foreign departments of the banks, and brokers whose specialist function is to bring together buyers and sellers among the dealers. Dealers buy and sell in the U.K. market only through the brokers and transactions are done solely by means of telephone. There are private lines linking banks to brokers.

There are many important foreign exchange markets in the principal cities overseas such as Paris, Zurich, Frankfurt, New York and Tokyo. Dealers in London can apply to these for rates and make a profit if the rate for a currency gets temporarily out of alignment between the two centres. This has the effect of speedily correcting inequalities and rates tend to be very close, whether quoted on the London market or in any of the cities mentioned above. There is a constant movement of rates due to the supply and demand of currencies, the variation in interest rates between the two centres, and other pressures such as threats of war, industrial action, or sanctions where, for example, oil producers arbitrarily limit supplies to their international customers. Dealers therefore look for their profits from the "spread" or margin between their buying and selling rates for each currency. They also make a profit by charging their customers exchange commission.

Prior to 1958 foreign exchange dealings were limited, but after 1958 many monetary restrictions and exchange controls were abolished and the foreign exchange markets began to expand. This growth was accelerated in the 1960s by the development of the Euro-dollar market. It was during this period that the foreign exchange department at the international bank grew in size from a small group of clerks to an establishment of around forty.

Dealers have a need for rapid, accurate and detailed information and for the rapid processing of deals. Computers therefore have a tremendous potential in foreign exchange [Hewitt 1976]. First, they can make information available to the dealer and enable him to deal more profitably and more confidently. Second, they can perform the back office work of automatically preparing confirmations of deals and telex schedules; reminding customers of when their deals mature; providing details of known movements over the bank's accounts with other banks (nostros) across the world; providing statistical information, and making interest calculations on loans and deposits.

A batch computer system was introduced into the foreign exchange department in 1964, with a change of machine in 1971 when decimalization was introduced and the existing machine was becoming obsolete. The principal input document was a copy of the deal ticket and this was used to post all entries, the only manual operation being the movement of funds by telex, cable or mail transfer. The batch computer system was introduced to help the bank undertake high volume routine processing in the foreign exchange field because management were worried that in the future it would not be possible to attract enough staff to carry the work out manually. At that time the bank's business was expanding rapidly. In human terms the batch system had a detrimental effect. The general opinion amongst staff was that it made work less interesting, although from the bank's point of view it greatly increased work accuracy. A change to an on-line/real time system was made in 1976.

9.2. ORGANIZATIONAL VALUES

Figure 1 shows that clerks, systems designers and user managers all saw the values of the bank as generally on the left hand side of the scale with a belief in shared values and a liking for standardized procedures and efficiency. The managers differed from the other two groups on discipline and task structure, believing that the bank did not want very strict discipline and favoured loosely defined jobs. The systems designers suggested that the bank placed rather more emphasis on the personal qualities of its staff than on efficiency and high production. All three groups moved over to the right when asked what they would like the values of the bank to be, although the two systems designers disagreed with each other on shared values and discipline and on the organization of work. The managers were much further to the right than either the clerks or the systems designers but this is not surprising as they played a major part in setting up the job satisfaction working party and stimulating the participative approach to the redesign of work. There was considerable divergence between the systems designers and the managers on whether the bank should use standardized methods or encourage employees to develop their own methods and use their own judgement, with both the systems designers in agreement on this point. This nicely illustrated two sets of values that were found in the bank where the O&M department, in particular, favoured standardized and tight job specifications and did not approve of the foreign exchange clerks designing their own work structure.

194

Figure 1. Perceived organizational values (modes)

Where systems designers and managers in each group mark different points on the scale, the point furthest towards the ends of the scale is the one shown.

Place a tick where you think this firm/office/department fits on the scale below

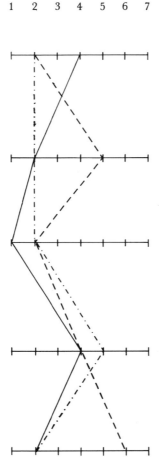

It has a strong belief in the importance of shared values – that employees should agree with its objectives and the way it carries out its operations

It likes to have disciplined employees who will put the firm's/office's interests first and willingly accept orders and instructions

It prefers to use standardized methods and procedures whenever possible as it believes that this assists efficiency in this kind of firm/office

It places a great deal of emphasis on efficiency and high production and less on personal qualities such as friendliness, trustworthiness, cooperation etc.

It feels the need to organize work activities into tightly structured jobs which are clearly defined and do not permit a great deal of individual discretion

A multiplicity of different values exists and is tolerated here. It is quite acceptable for employees to hold very different views on what the firm should be doing and how it should be doing it

It does not believe in very strict discipline and tries to provide a situation in which people can pursue their own interests

It likes employees to work out their own methods for doing things and to use their own judgement when taking decisions

It places a great deal of emphasis on personal qualities such as friendliness, trustworthiness, cooperation etc. and less on efficiency and high production

It tries to organize work in such a way that employees have loosely defined and structured jobs which permit a great deal of individual discretion

———————— N = 36   Clerks
– – – – – – N = 2   Managers
– · – · – · – N = 2   Systems designers

195

Figure 2. Desired organizational values (modes)

Please tick where you *would like* this firm/office/department to fit on the scale below.

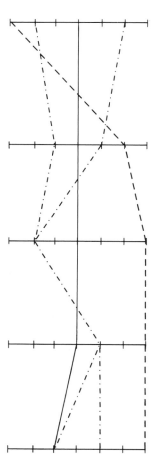

It should have a strong belief in the importance of shared values – that employees should agree with its objectives and the way it carries out its operations

It should try to create disciplined employees who will put the firm's/office's interests first and willingly accept orders and instructions

It should try to use standardized methods and procedures whenever possible as this would assist efficiency in this kind of firm/office

It should place a great deal of emphasis on efficiency and high production and less on personal qualities such as friendliness, trustworthiness, co-operation etc.

It should try to organize work activities into tightly structured jobs which are clearly defined and do not permit a great deal of individual discretion

A multiplicity of different values should exist and be tolerated here. It should be quite acceptable for employees to hold very different views on what the firm should be doing and how it should be doing it

It should have an easy-going kind of discipline and try to provide a situation in which people can pursue their own interests

It should encourage employees to work out their own methods for doing things and use their own judgement when taking decisions

It should place a great deal of emphasis on personal qualities such as friendliness, trustworthiness, cooperation etc. and less on efficiency and high production

It should try to organize in such a way that employees have loosely defined and structured jobs which permit a great deal of individual discretion

——————— N = 36 Clerks
– – – – – – N = 2 Managers
· – · · – · · – · N = 2 Managers

196

The senior manager saw the bank's philosophy as trying to create a family and team spirit in a situation which did not require imposed discipline and where there was a great deal of joint discussion. He believed in accountability but wanted all staff to feel that they were involved and vital members of an integrated team, regardless of seniority. The departmental manager's views were similar. He wished to have employees who thought for themselves, showed initiative, liked their jobs and were hard-working. He pointed out that a major difference between banking and other industries was that banking had to have tight controls because they were handling other people's money. This led to the bank formulating explicit rules which could not be diverged from, and this could make it difficult to foster independence of mind and encourage original thought.

The answers to the question on "the best form of department structure for clerical staff in general" also show considerable similarity of view between the three groups (Figure 3). The clerks favoured a considerable amount of self-determination although they preferred targets to be set and monitored by supervision and both managers and systems designers agreed with them on this. There was a strong belief that everyone should have access to all the information they required and one of the two systems designers had a greater preference for well-defined jobs and close supervision than the other. The senior manager pointed out that his answers represented what he was trying to achieve in terms of departmental structure, and that this was the reason for the job satisfaction exercise. In his opinion the organization of much of the work was too similar to factory-type production.

The systems designer on the right of the scale was responsible for the design of the on-line system in foreign exchange and his views were similar to those of the two managers. He believed that the bank should encourage independence of ideas and that if standardized methods had to be used then they should be developed with the agreement and cooperation of the employees responsible for the work. He also believed that employees could be trained to do a variety of jobs, in this way increasing both their enjoyment and their efficiency, and that discipline should be imposed by the work group rather than the supervisor. Opinion on the "general characteristics of the majority of staff in the foreign exchange department" was somewhat surprising with the clerks being rather cautious about their own abilities, management rating their self-management skills highly and the systems designers generally on the left hand side of the scale, seeing the clerks as preferring a structured work environment without too many opportunities for discretion (Figure 4).

The impression given by this analysis of organizational values is that all three groups regarded a stimulating and flexible work environment as desirable and that management, in particular, was in favour of this. Group values and the climate of opinion therefore fitted a philosophy of participation and efforts to increase job satisfaction.

Figure 3. Organizational models (modes)

Please tick on the scale below what *you* believe to be the best form of department structure for clerical staff in general.

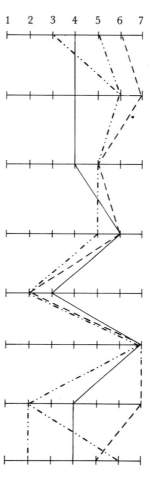

Jobs should be clearly defined, structured and stable	Jobs should be flexible and permit group problem solving
There should be a clear hierarchy of authority with the man at the top carrying ultimate responsibility for all aspects of work	There should be a delegation of authority and responsibility to those doing the job regardless of formal title and status
The most important motivators should be financial, e.g., high earnings and cash bonuses	The most important motivators should be non-financial, e.g., work challenge, opportunity for team work
Jobs should be carefully defined by O&M department, management services or supervision and adhered to	The development of job methods should be left to the group and individual doing the job
Targets should be set by supervision and monitored by supervision	Targets should be left to the employee groups to set and monitor
Groups and individuals should be given the specific information they need to do the job but no more	Everyone should have access to *all* information which they regard as relevant to their work
Decisions on what is to be done and how it is to be done should be left entirely to management	Decisions should be arrived at through group discussions involving all employees
There should be close supervision, tight controls and well maintained discipline	There should be loose supervision, few controls and a reliance on employee self-discipline

———————	N = 36	Clerks
– – – – –	N = 2	Managers
··—··—··—	N = 2	Systems designers

198

Figure 4. Models of man (modes)

Please tick on the scale below how you see the general characteristics of the majority of staff in this department.

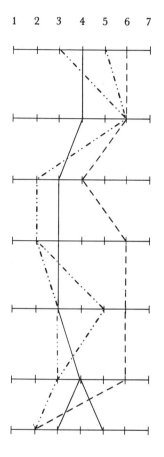

	1 2 3 4 5 6 7	
They Work best on simple, routine work that makes few demands of them		They Respond well to varied, challenging work requiring knowledge and skill
They are Not too concerned about having social contact at work		Regard opportunities for social contact at work as important
They Work best if time and quality targets are set for them by supervision		Are able to set their own time and quality targets
Work best if their output and quality standards are clearly monitored by supervision		Could be given complete control over outputs and quality standards
Like to be told what to do next and how to do it		Can organize the sequence of their work and choose the best methods themselves
Do not want to use a great deal of initiative or take decisions		Like, and are competent to use initiative and take decisions
Work best on jobs with a short task cycle		Able to carry out complex jobs which have a long time span between start and finish

——————— N = 36 Clerks
– – – – – – N = 2 Managers
– ·· – ·· – ·· – N = 2 Systems designers

199

9.3. WORK VALUES

Values in a work situation will influence attitudes to such important things as efficiency, happiness and a sense of fairness and justice, and so the managers and systems designers were asked to describe what they saw as the principal factors contributing to these. Their answers were then categorized in terms of rational factors (cognitive), emotional factors associated with pleasure or displeasure (cathectic) and factors associated with equity (moral). These are shown in Figure 5.

Figure 5. Work values (systems designers' and managers' view)

EFFICIENCY

Role expectations
(what is expected of staff)

Cognitive	**Cathectic**	**Moral**
An interest in knowing the the job	Motivation	
Understanding the purpose of the job	A happy team spirit	

Role requirements
(what the organization must provide)

Cognitive	**Cathectic**	**Moral**
An understanding of organizational objectives*	An understanding management	
Well-defined organizational structure*	A happy environment*	
Clear job definitions*		
Sensible work methods*		
Written rules of procedure*		
Responsibility*		
Interesting work*		
An efficient management		

HAPPINESS

Role expectations
(what is expected of staff)

Cognitive	**Cathectic**	**Moral**
	Staff who like their work	
	Trust and respect*	
	Job satisfaction	

Cognitive	Cathectic	Moral
Sufficient responsibility*	Good working environment	Good promotion opportuni-
Interesting work*	Good conditions of work*	ties
Varied work*	Good relationships between	Satisfactory pay
Involvement in job design*	management and staff	Good prospects*
	Management must consider	Fairness*
	and encourage all employees	

FAIRNESS AND JUSTICE

Role expectations
(what is expected of staff)

Cognitive	Cathectic	Moral
—	—	—

Role requirements
(what the organization must provide)

Cognitive	Cathectic	Moral
	Managers who always listen	Explanations for decisions
	to the problems of staff	on promotion
	Managers who take staff into	Equal rewards for jobs of
	their confidence	equal responsibility*
		Allocation of work to those
		most competent to do it*
		Straight speaking
		Justice is seen to be done

* indicates that this answer has been given by a systems designer.

These replies are interesting for two reasons. First, management perceived both efficiency and happiness to be related to certain employee attitudes, both intellectual and emotional. Second, a majority of the management answers fell in the cathectic category. Not only must employees bring positive attitudes to the work situation, management must itself act and react in a way that will create these attitudes, and this action should be in terms of developing the kinds of personal relationships and work environment that will cause staff to respond with pleasure to the work situation. In contrast the systems designers saw efficiency and happiness as more related to a rational work environment, structured in such a way as to produce efficiency and satisfaction. Both groups saw a sense of fairness and justice as resulting from equitable personnel policies on such things as pay, promotion and the distribution of work, and from such policies being made explicit.

Because efficiency, happiness and a sense of fairness and justice are believed to be components of job satisfaction, managers and systems designers were asked to define this term. Management definitions were,

—To enjoy coming to work and to go home with the feeling that the job was worth doing.

—When a person goes to work in the morning and actually looks forward to doing a good day's work. Although people cannot be expected to enjoy every day, on average they should see their daily task as a pleasant event.

One of the systems designers defined it as,

1. Doing a job which is of benefit to someone;
2. Doing a job which stretches my capabilities;
3. Doing a job which is well rewarded.

An attempt to obtain an insight into the values associated with the design of computer systems in the bank was made by asking the systems designers and the user manager to describe and rank in order of importance the set of activities that they would include in the phrase "systems analysis".

The manager's list was phrased in terms of objectives of value to a user.

To study:
1. Reliability;
2. Flexibility;
3. Accuracy;
4. Job enrichment;
5. Environment;
6. Staff reduction.

Only one of the systems designers answered the question and he provided a design task sequence.

1. The definition of objectives;
2. Analysis of available information;
3. Planning system alternatives;
4. Testing the practicability of ideas on the user;
5. Documentation.

The systems designers were also asked if they used any basic design principles when designing the computer systems. The designer of the bank's foreign exchange batch system saw his role as reducing clerical effort, providing more accurate information, providing a set of information and processing facilities which would not be available with a manual system and capturing data as near to the source as possible.

9.4. GOAL SETTING AND ATTAINMENT

The existing batch computer system in the foreign exchange department had been introduced when the bank was expanding rapidly and management were anxious that they might not be able to attract enough staff to process the work manually. It carried out a variety of tasks for the exchange department clerks, recording each individual transaction carried out by dealers so that the bank's accounts and books were always up to date; providing basic documents for customers who had bought or sold currency; generating diary notes for the clerks to remind them of actions that had to be taken, and providing management with information about the business which assisted the determination of future strategy. It had the advantage of using one document, the deal ticket, as the basis of all other bank documents.

The technical objective of the proposed on-line system was to further streamline operations and to provide a faster and more up-to-date management information service, part of which would be in real time mode. An important business objective was to provide a faster and more efficient service to the bank's customers through sending rapid confirmation of deals. The system also had an important human objective of providing an opportunity for improving the organization of work and thus the job satisfaction of the bank clerks. It was hoped that this would have the practical result of reducing labour turnover in the exchange department. The system had to meet the needs of the dealing room, where there were few job satisfaction problems but a requirement for fast and accurate information; and of the clerks in the exchange department who were suffering from a clack of job interest because of the extreme segmentation of work.

One problem associated with the introduction of the system was related to costs. The on-line system would be expensive to introduce, yet in 1980 it was proposed that the international bank should give up its own computer and transfer to the computer of the clearing bank with which it was associated. This could mean that the new system would only just be showing a profit when it had to transfer to a different computer. This led to considerable discussion on whether the system should be on-line and real time or on-line only. Pressure from the dealers influenced the choice of real time.

The first proposal for the on-line system was prepared by the computer services department who had been carrying out some investigations in the bank to identify where an on-line system would prove most valuable and which department should be given highest priority with such a system. The choice was foreign exchange and the management and staff of that department played an important role in the design processes from the outset. A feasibility study was prepared by computer services in which a number of technical alternatives were considered. These were, first, to have an on-line/real time system, second, to have on-line with batch processing, and third to leave the current batch system as it was. The systems designers' view at this stage was that both the first and second alternatives were viable but that the third was not.

There were a number of technical constraints associated with the design of the system. A major constraint was the processing capacity of the computer. Data processed in batch mode overnight could not be allowed to run into the following morning or it would prevent the on-line system from coming into operation. Another constraint was a requirement to purchase Burroughs terminals as it was bank policy to link the main frame to peripheral equipment of the same manufacturer.[1]

---

1. The bank already had two Burroughs 3500 computers.

Business constraints were associated with the need for a strict adherence to bank rules in the design of the system. There had to be double checking; documents must be signed by those authorized to sign them and bookkeeping procedures must be adhered to. There were also time constraints, for the system had to be ready by January 1976 if it was to be cost-justified before the change-over to the clearing bank computer.

9.4.1. Design Alternatives

When the batch system was designed in 1964 two alternative approaches were considered. The first was to introduce a batch system to handle the bank's operational work and later to convert this to a system which could produce information for management. The second alternative was to design a combined operational and information system at the start. This would give management better control over day-to-day problems and also better information by which to make future planning decisions. Two problems inhibited the adoption of this second approach. The first was that it would take a considerable time to design such a system and the bank wanted to install its computer system fairly quickly. The second was that management at that time did not have a great deal of experience in using computer-produced information. It was therefore decided to use a batch system to handle the bank's daily processing and later evolve this into an MIS system. This meant that the design of the system was influenced by the future objective of producing information quickly and accurately.

No human objectives were set for this system and the system designer responsible for developing the system said that such things were not thought of in the 1960s. The user had no experience of computers and therefore made no demands and the system had little effect on the organization of the office for it merely computerized the manual system.

With the on-line system the technical design alternatives considered included the three identified in the feasibility study, together with one other, the purchase of a package system for money market activities. Such a package existed although it would not have met the needs of the bank exactly. The systems designer evaluated the pros and cons of each alternative as shown in Table 1.

Alternative 1 was eventually chosen although the system was not designed to operate in total real time. Dealers' information was updated immediately but bank information was processed overnight.

The proposal was for important data on deals – the customer's name, the rate, date of deal, currency and amount – to be put into the computer immediately the deal was made. This would produce an immediate update of customer purchases. It would also provide information for the bank on how much currency was bought and sold over a particular period. Other data fed into the computer would be usedto assist the preparation of advices to customers and other banks. This would be put in around 20 minutes after the deal was completed, its accuracy would be checked via the terminals and the computer would then produce the necessary documentation. The VDUs would also provide an enquiry facility on the cus-

Table 1. Design alternatives

Alternative	Advantages	Disadvantages
1. On-line/real time + enquiry facility	1. Improved information 2. Speed of output 3. Saving in paper 4. Opportunity for dept. reorganization	Cost Complexity Few
2. On-line/batch + enquiry facility	2. and 3. above A big saving in the costs of producing annual audits, etc.	Cost Few
3. Leave existing batch system	Cost	It was not meeting requirements. Dept. organization could not be improved
4. Buy package system for money market	Lower cost and other advantages listed above	Would not meet all the requirements of the international bank Possible maintenance problems

tomer's position, details of his deals and other information of importance to the bank. The aim would be to answer all enquiries through the VDUs. The principal factor influencing the choice of technical solution was the bank's desire to ensure a strict control over the affairs of its customers. Two German banks had recently collapsed and so the bank was anxious to adopt the most efficient form of control.

Human design alternatives were the responsibility of the specially set up job satisfaction working party and the technical design group played no major part in formulating these, although they accepted that they must not create a system which imposed any major constraints on the way in which foreign exchange organized its work. To facilitate the user's reorganization task they therefore kept the computer system as flexible as possible.

9.5. ADAPTATION

The adaptation of the foreign exchange clerks to the on-line system was greatly facilitated by the fact that in assuming responsibility for the redesign of work, they had a considerable amount of control over the processes of adaptation. Participation of this kind not only enabled them to identify and solve their own efficiency and job satisfaction problems, it also provided them with a learning experience. In analyzing the basic functions of their department they gained a greater understanding of the nature of the problems the working party had been set up to solve; in thinking through alternative ways of reorganization and fitting these to the technical potential of the computer they also learnt a great deal about how the new computer system could be used.

The approach adopted by the international bank was what the author calls "representative democracy". With this approach a design group is formed which is representative of all grades of staff in the user department, and, if a new computer system is being introduced, may also include the technical systems designer. The departmental manager may or may not be a member of this design team, depending on his own wishes. This approach was pioneered by Professor Louis Davis, Director of the Quality of Working Life Centre at the University of California, Los Angeles. His personal philosophy has always been that *"no-one has the right to design another person's work situation; the role of the expert is to assist the employee to design hiw own"*. The international bank believed in this philosophy, subject to such an approach fitting with the bank's objectives.

The job satisfaction working party had been organized before the author became involved in the project and consisted of six representatives of the bank clerks drawn from different levels and functions in the foreign exchange department.[2] Five represented the exchange department but the group also included a dealer to ensure that relationships between the dealing room and the exchange department were carefully considered. One member of the group had considerable experience of the technical aspects of computing and he had the additional role of liaising between the working party and the systems designer. Although the working party kept in constant touch with the systems designer, he did not join the meetings on a regular basis. The head of the exchange department also decided not to join the working party although he too kept in close touch with its activities. The committee began to meet in December 1974 and made its report to senior management at the end of May 1975.

One of the first tasks of the job satisfaction working party was to define its objectives and it decided that its principal aim was to develop a new form of work organization for the exchange department that increased efficiency and led to more job satisfaction through the provision of larger and more challenging jobs. The existing structure of the department was based on the different functions associated with the processing of deals and produced a great deal of work segmentation. Also the introduction of the batch computer system had brought with it a number of boring and routine jobs, for example the coding of input vouchers. It had also virtually eliminated ledger work which previously had been a source of satisfaction to many bank staff. Working party members recognized that by changing departmental structure they were also going to change drastically many people's jobs. They therefore decided that they must also develop strategies for training, career development and flexible hours. Again, they saw their task as not merely concerned with designing the human part of the new work system but also

---

2. This group were democratically elected by their colleagues.

with its implementation and therefore an implementation strategy would be required.

The working party also had to find answers to a number of problems related to efficiency. The department had to deal with a large number of currencies, but dollars were the dominant currency and these tended to overwhelm the rest of the work. There was also a problem of work peaking. Some dealers had two work peaks in a day, 8.00 a.m. to 10.00 a.m. and late afternoon and these affected the work of the exchange department. Any new work system must also provide for the tight controls that banking required. This could present a design problem as it is not easy to associate a strict control system with staff opportunities for the use of discretion and creative thinking. Management also had a number of objectives that had to be catered for in any redesign. The head of foreign exchange hoped that an improved work system, allied to an enhanced technology, would enable him to reduce staff requirements by around twenty, as well as improving the efficiency of the office by increasing the speed of throughput of work and reducing the number of errors. He was also anxious to reduce staff labour turnover, which was high at this time, particularly amongst experienced clerks of section head level, most of whom were in their early and middle thirties. This movement was due to a lack of promotion prospects within the department and to the attraction of higher salaries elsewhere.[3] However, the fact that the exchange department had a high labour turnover was of assistance here as it meant that a saving in staff could be achieved without causing any personal distress.

9.5.1. The Batch System Structure of Work

A report prepared by the manager of the foreign exchange department provides a brief history of organizational changes in the department. It describes:

1. *The "old" department*
   Prior to November 1971, the foreign exchange department dealt with a great variety of the bank's transactions in foreign currencies. Roughly the operations could be divided into (*a*) exchange and money market operations; (*b*) customer services.

   The processing of the transactions under (*a*) involved: (*i*) exchange operations, effected through the international exchange market (inter-bank deals); (*ii*) deposits and loans, effected through the international money market (inter-bank deposits/loans).

   The processing of the transactions under (*b*) involved: (*i*) exchange operations, customer-based; (*ii*) deposits and loans, customer-based; (*iii*) issuing of currency drafts; (*iv*) effecting mail and telegraphic transfers on behalf of customers; (*v*) operating current and sight accounts in foreign currencies; (*vi*) customer's account documentation, etc.

---

3. A report of the manager of the department states that in 1973 turnover was nearly 50%. A third of these gave as their reason for leaving an absence of job satisfaction.

The processing of inter-bank exchange operations and deposits/loans, (a) above, could be classed as a stereotype operation. The counterparty, paying and receiving agent, amounts and rates may differ, but the procedure is the same. It could be described as an arduous and often boring and repetitive job. At the same time it is a job which demands the utmost concentration and efficiency, for any mistake in any one of the details may result in considerable monetary loss to the bank.

The processing of the operations mentioned under (b) on the other hand are of a far more varied nature. It gives scope for a variety of activities, such as: (i) personal or written contact with customers; (ii) finding the most effective route for payments abroad; (iii) handling the queries (or complaints) from customers; (iv) obtaining documentation, such as signature cards, joint and several accounts forms, etc.; (v) handling of requests for customers wishing to cash sterling against their currency holdings, and a thousand and one other services which the bank is required to perform for its customers.

2. The "new" department

In November 1971 most of the non-market operations were transferred to a new department, called accounts and payments department, while the market operations, i.e., operations effected in the dealing room were to be handled by the present exchange department, together with customer-based currency fixed deposits and loans and customer-based call and notice accounts. At the same time sterling market operations previously handled by the then cash department, were now to be processed in the exchange department.

Thus a large proportion of the repetitive operations remained in the exchange department, whereas the majority of the varied operations went to the newly formed accounts and payments department. It is true to say that some customer-oriented operations remained within the department, but the proportion of these operations to the more stereotyped operations became much smaller. Figure 6 shows the organization of the department in 1975. Figures 7 and 8 provide examples of the work of a section and of the very routine nature of lower grade jobs.

In 1975, at the time of this research, the exchange department had a very long grading structure as shown below.

Head of department (Principal)	
Assistant principals	
Supervisors	Management
Section heads	
Deputy section heads	
Checkers	
Coders	
Typists	
Filing clerks	

Each of these jobs had a different grade ranging from grade 1 filing clerks to grade 7 assistant principal. The non-management group encompassed responsible and challenging jobs at the top of the hierarchy to extremely small and rou-

Figure 6. International bank. Pre-change organization of exchange department (total staff = 68)

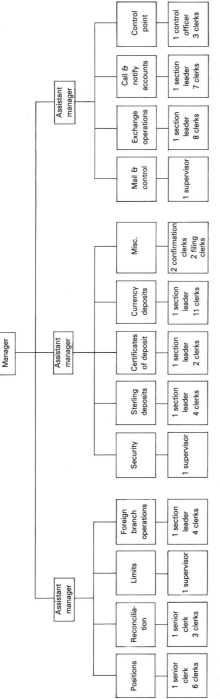

Positions
These clerks work with the dealers. They book deals and give deal tickets to the appropriate section.
They also keep a record of funds going in or out (the position) and know the value date of deals.

Reconciliation
All statements and ledgers are held here and reconciled with one another. If they do not match the query is sent to the appropriate section leader for investigation.

Limits
This is a check that the volume of deals with other banks is within the limits set by the international bank.

Foreign branch operations
Handles instructions from foreign branches.
a. Takes deposits or loans from customers and sends customers contracts.
b. Does branch reconciliations.
c. Does deposit guarantees.

Sterling Deposits
Handles deals concerning sterling, keeps sterling record.

Certificates of deposit
Formal documentation that is associated with a loan. Customer records.

Currency deposits
This section is concerned with dealing with other banks. Borrowing and lending money to the customers of these banks who act as agents for their customers.

Miscellaneous
a. Clerks who handle confirmations produced by the computer indicating that a deal has gone through.
b. Clerks who file documents.

Mail and control
Final check before mail goes out.

Exchange operations
Responsible for processing all foreign exchange transactions. Exchange transactions deal ticket is main input.
Handles reconciliation problems.
Prepares notes for dealers on sterling position each morning.

Call and notify accounts
Deposits which are not made for a fixed period of time. The customer wants notification when rates change so that a better rate of interest may be obtained.

Control point
This section is the department interface with the computer. Distributes computer printouts.
a. Daily balance list.
b. Diary statements (record of deposits). Handles computer input.

209

Figure 7. Currency fixed deposits and loans section

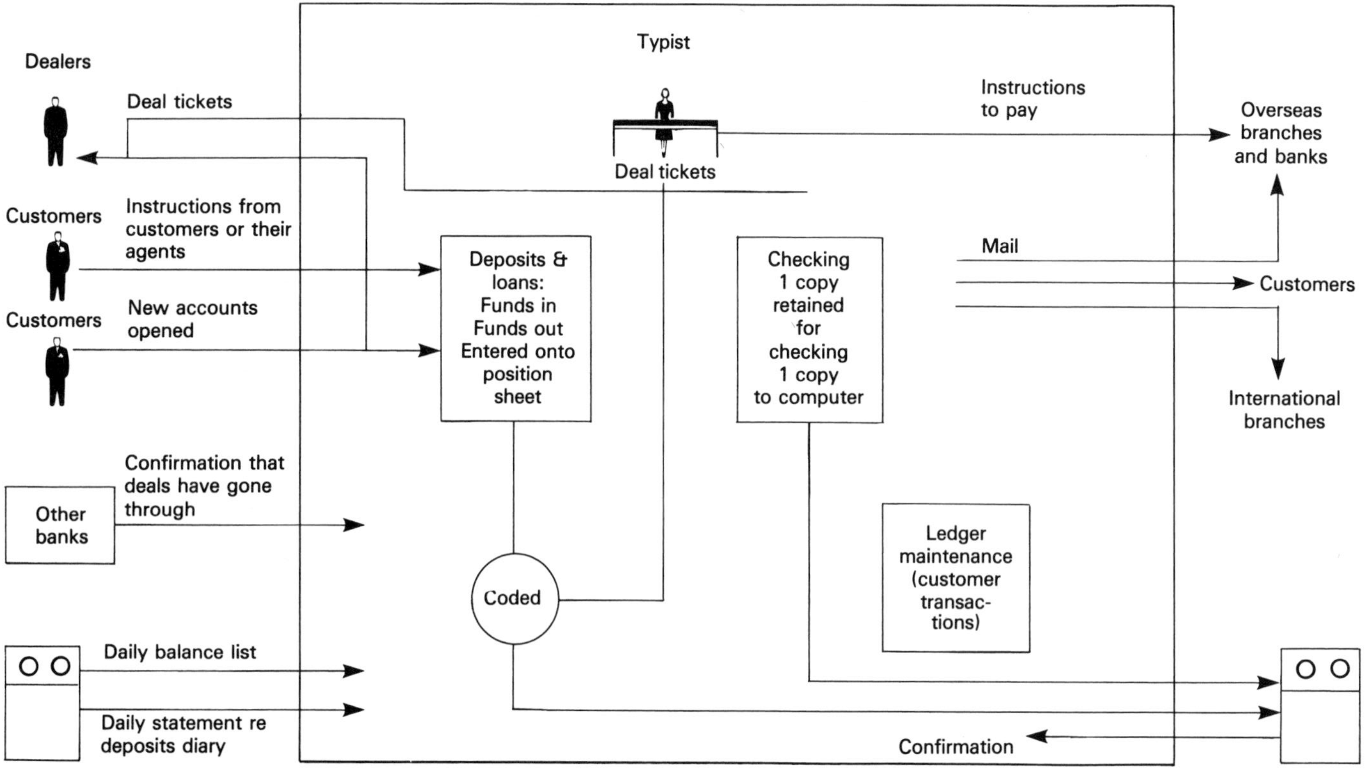

210

Figure 8. Examples of routine jobs

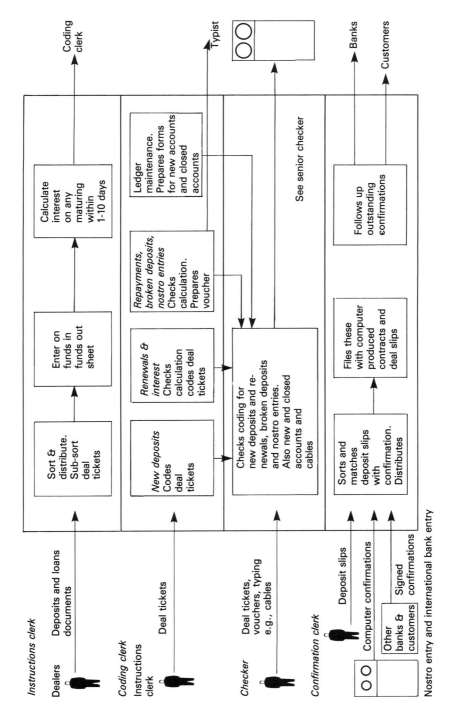

211

tine jobs at the bottom. For example, a deputy section head in exchange operations had a set of tasks with a considerable problem solving component. He or she was responsible for identifying the reasons for discrepancies between statements and ledgers, a complex task that could take up to an hour per query. In addition, there was responsibility for checking cables and banker's payments and handling extensions or cancellations of deal options. In contrast, the checker's job, as the name implied, was solely concerned with checking the work of other clerks. The checker checked the codes on deal tickets, sorted and matched confirmations of deals with deal tickets and did other checking operations. A coder coded the deal tickets and checked calculations prior to the input of data into the computer. In addition there were instructions clerks who handled the deal tickets when they arrived in the department from the dealing room and confirmation clerks who handled computer-produced confirmations of deals.

Table 2 compares the structures of these jobs in terms of what are considered to be good job design criteria.

One of the roles of the author was to carry out a survey of job satisfaction in the exchange department to assist the working party in identifying job satisfaction needs. The results are shown in Table 3.

This Table gives the negative ends of the job satisfaction scales only. The working party received from the author the full job satisfaction analysis. These results show considerable work dependency, a feeling of working under pressure, a desire for larger jobs with the opportunity to see a piece of work through from start to finish, a wish to use more judgement and take more decisions and a desire for more challenge and less routine. These results are related to different aspects of work. General measures of job satisfaction are shown in Table 4.

Less than a third of staff had very positive feelings towards their work and around a quarter had very negative feelings, confirming management's belief that job satisfaction in the department could be improved. Views on the batch computer system were mixed. 56% of clerks said that it increased their efficiency but 36% said that it made no difference. 24% said that it increased the interest of their work, 20% said that it decreased this and 56% said that it made no difference. Discrepancies between job needs and job requirements arose from the fact that, although a great deal of work was routine, the exchange department staff were above average in intelligence and above average in educational attainment. The author was struck by the fact that very intelligent young people who had succeeded in gaining good educational qualifications in school were brought into the bank and given the extremely routine and simple jobs of coder or checker.

Using the job satisfaction survey as a guide, the design team gave its attention to a number of alternative ways of reorganizing the department. Three alternatives seemed worth considering:

1. Maintaining the existing functional structure;
2. Eliminating functional divisions and organizing the department around the logical work flow;
3. Dividing up the department into groups based on currencies.

212

Table 2. International bank. Job structure comparisons

Control levels and functional groupings (HIGH CONTROL LEVELS)

Control / functional level	Reduce boredom → descriptors
LEVEL 1 – OPERATIONAL LEVEL	
Variety	Give sense of personal control (of methods, of work sequence)
Means choice	Give feeling of achievement (Use of judgement)
Knowledge choice	Give sense of making important contribution (Use of initiative; Clearly defined start to job; Clearly defined end to job; Uninterrupted task sequence; Long task cycle 20+ minutes)
Job identity	Give sense of team work (Visible contribution to product or service; Works as member of group)
Job relationships	Give sense of confidence (Considerable inter-action required; Clear work objectives)
LEVEL 2 Goal clarity	Problem prevention (Objectives not too easy or too difficult; Can requisition required resources)
Anti-oscillation	Efficiency improvement (Can correct errors, solve problems; Coordinates own work activities)
LEVEL 3 Optimisation	Creativity (Coordinates group work activities; Can improve methods)
LEVEL 4 Development	Autonomy (Can improve product or service; Individual free from supervisory control)
LEVEL 5 Overall control	Key task (Group free from supervisory control)

Definition of skills:
A. Communicating in writing
B. Communicating verbally
C. Arithmetical skills
D. Machine operating skills
E. Checking/monitoring/correcting
F. Problem solving
G. Coordinating
H. Supervising

"These are the day to day operating tasks"

Foreign exchange batch system

Job	Skills	No. of tasks	No. of skills	of methods	of work seq.	Use of judgement	Use of initiative	Clearly defined start	Clearly defined end	Uninterrupted task seq.	Long task cycle 20+ min	Visible contribution	Works as member of group	Considerable inter-action	Clear work objectives	Objectives not too easy/difficult	Can requisition resources	Can correct errors	Coordinates own activities	Coordinates group activities	Can improve methods	Can improve product/service	Individual free from superv.	Group free from superv.	Key task
Deputy section head – exchange operations	✓	L	5 A/B/E F/H	✓	✓	✓	✓	✓	✓		with some cases	✓	✓	✓	✓	✓	✓	✓	✓						Problem solving / Checking
Instructions clerk – exchange operations	✓	M		✓			✓	✓	✓	with some cases	with some cases	✓	✓	✓	✓	✓	minor								Collecting inf., calculating, recording
Checker exchange operations	✓	S	1 E	✓			✓	✓	✓			✓			✓										Checking
Senior checker currency fixed deposits	✓	S	2 A/E	✓			✓	✓	✓			✓			✓										Checking
Coding clerk currency fixed deposits	✓	S	2 A/E	✓			✓	✓	✓			✓			✓										Coding / Checking
Confirmation clerk currency fixed deposits	✓	S	2 A/E	✓			✓	✓	✓			✓			✓										Checking, filing, collecting information

213

Table 3. Opinion on existing job structure
(in percentages)

Category	Sub-category	FACT/PREF	%	Statement
VARIETY	Routine and challenge	FACT	19	Work is usually or always routine
VARIETY	Routine and challenge	PREF	48	There is too little challenge
VARIETY	Routine and challenge	PREF	67	There is too much routine
DISCRETION	Means choice	FACT	31	Almost everything has rules or set procedures
DISCRETION	Means choice	FACT	31	Can hardly ever choose our work methods
DISCRETION	Means choice	FACT	17	Can hardly ever choose work sequence
DISCRETION	Means choice	FACT	19	Can hardly ever choose next task
DISCRETION	Means choice	PREF	31	Would like more choice of work methods
DISCRETION	Means choice	PREF	20	Would like more choice of work sequence
DISCRETION	Means choice	PREF	30	Would like more choice of next task
DISCRETION	Knowledge choice	FACT	36	Can hardly ever make decisions, use judgement
DISCRETION	Knowledge choice	PREF	57	Would like more opportunities to make decisions, use judgement
JOB IDENTITY	Task cycle and integration	FACT	42	Has very little or little chance to see a piece of work through from start to finish
JOB IDENTITY	Task cycle and integration	PREF	54	Would like more chance to do this
JOB RELATIONS	Work dependency	FACT	70	Inefficient work by others can almost always or usually, hinder me
JOB RELATIONS	Work dependency	FACT	83	Inefficient work by me can almost always or usually hinder others
JOB RELATIONS	Work dependency	PREF	31	I would like there to be less work dependency
GOALS	Deadlines	FACT	70	There is too little or barely enough time to do work

N = 36

214

Table 4. Positive and negative feelings about work
(in percentages)

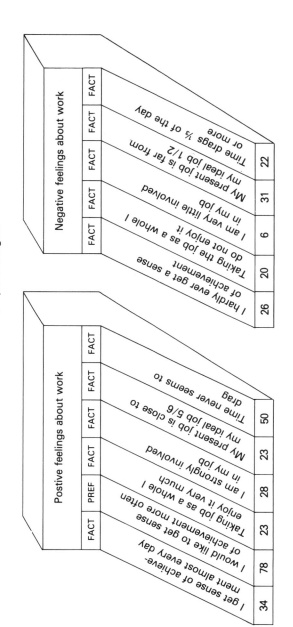

Negative feelings about work

	FACT	FACT	FACT	FACT	FACT
I hardly ever get a sense of achievement	Taking the job as a whole I do not enjoy it	I am very little involved in my job	My present job is far from my ideal job 1/2	Time drags ⅔ of the day or more
26	20	6	31	22

Positive feelings about work

	FACT	PREF	FACT	FACT	FACT	FACT
I get sense of achieve-ment almost every day	I would like to get sense of achievement more often	Taking job as a whole I enjoy it very much	I am strongly involved in my job	My present job is close to my ideal job 5/6	Time never seems to drag
34	78	23	28	23	50

The first alternative had the advantage of minimizing disturbance but had the disadvantage that it did nothing to improve job satisfaction. The working party also saw it as poor in meeting the staff needs for work variety. It was also inflexible and hindered staff mobility. If one section was very busy staff in other sections would not have the skills to help.

The second alternative would produce an improved work flow but would be very difficult to control and could lead to staff dissatisfaction as there would be no groups with which they could identify.

The third alternative, division by currency, was the one chosen. The dealing room was already divided by currency blocks and it seemed logical to divide the exchange department in the same way. This would give job variety to staff, since within each currency block all types of transactions would be processed. It would also give a sense of group identity to the members of each currency section. Also the supervision and control of such a system would be relatively easy.

Having selected a viable alternative the working party next had to set out the detail of this form of organization. The most efficient currency breakdown had to be worked out, based on the volume per currency transaction, so that the new sections would have similar work loads of one or more currencies. Details of the organization of work within each section had also to be formulated and it was decided that a multi-skilled group structure would be the most appropriate, with each member eventually trained to do all the jobs of the section, from input through the final check before documents were despatched. This would increase the interest of work and also enable a clerk to start a deal and effectively monitor and complete its passage through the system. This approach would also increase the clerk's knowledge, experience and responsibility. The working party suggested that once the system was introduced job evaluation should be based on knowledge and experience and not on grade as at present.

The VDU system would provide clerks with the opportunity to input their own deal information and transmit the deal to the computer. When the confirmation was produced this could be handed, with the deal ticket, to the section's signing officer who would check the deal and then sign it. It was thought that this approach woudl reduce the processing time for deals while allowing checks to be still retained. This, together with the elimination of coding and checking by the VDUs, would enable the department to achieve a staff saving of 17. Figure 9 shows the task structure for each currency section with the proposed system of work.

There would be four currency sections – dollars, sterling, Swiss franc/guilder/sundry, Deutschemark/French francs, and a service section. Each section would require a staff of ten. The available mix of tasks comprised:

a. Tasks associated with computer input. Collecting information, adding to information, inputting information via the VDUs, checking the accuracy of input. When necessary establishing the integrity of input by using the VDU enquiry facility to check on customer status.

b. Tasks concerned with real-time computer output. Dealing with confirma-

Figure 9. The international bank. Task structure for each currency section after introduction of new computer system
(deals processed in real-time mode; bank information processed in batch mode)

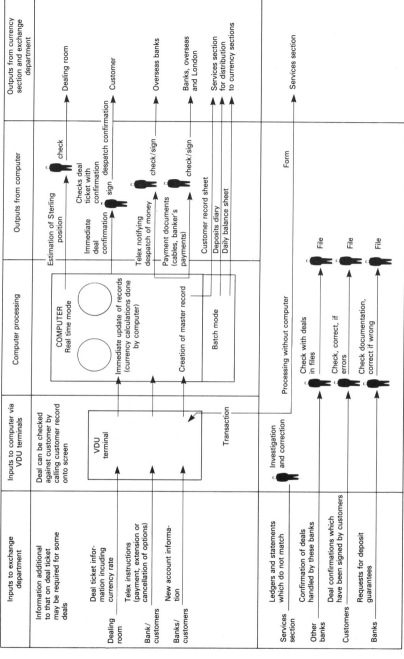

tions and payment documents such as cables and banker's orders. Checking the correctness of these before despatch.

c. Tasks associated with batch computer output. Identifying the reason for errors in printed output documents, balance lists, etc.

d. Tasks still independent of the computer. Information coming into foreign exchange department from other banks, e.g., confirmation of deals handled by other banks, requests for deposit guarantees, etc. Problems associated with meeting customer instructions, reconciling documents, correcting errors.

e. Control procedures such as final check and signing before documents are despatched.

Staff within a section would have to deal with the three basic tasks of exchange operations, fixed deposits/loans and sight and notice accounts, previously handled in separate sections.

The working party also made recommendations about the environment of the department, training, and implementation and designed a floor plan showing the layout of the currency sections. They also recommended that they should remain in being once the new structure was implemented so that they could monitor it and iron out problems. The working party's report was presented to the bank management at the end of May 1975, and was accepted.

9.5.2. Other Adaptation Structures and Processes

The job satisfaction working party assumed complete responsibility for the design of the human part of the new work system but there were also other committees which played a role in the design processes and helped the change from one level of technology to another. A progress group was formed consisting of representatives from EDP, O&M, and foreign exchange, together with the bank inspector. This group met every three weeks to discuss any problems that were being encountered with the design of the technical part of the system and provided a link between foreign exchange management and the EDP department. The EDP group also met every two weeks to discuss progress and there was an on-line subcommittee to consider the general use of on-line systems throughout the bank. The members of this included EDP and representatives from every area of the bank.

The foreign exchange department in the international bank therefore used the introduction of a new computer system to provide an opportunity for rethinking organizational needs and for developing a socio-technical system that would decrease costs, improve efficiency and increase job satisfaction. The approach it used to achieve these goals, the giving of responsibility for the design of the new structure to the staff who would have to operate it, was a great aid to the successful assimilation of the system into the department. The job satisfaction working party understood very well the needs of their colleagues in the department, and were familiar enough with foreign exchange work to be able to evolve alternative forms of organization structure and to make an optimal choice. The initia-

218

tive for this participative approach had initially come from senior management but the manager of the foreign exchange department and the designer of the on-line system were also very committed to it. The systems designer's argument was that foreign exchange was a very complex activity and that the knowledge of the user was far greater than that of the computer specialist, therefore the user should have responsibility for the non-technical parts of the design process.

9.6. INTEGRATION

In December 1975 the department was reorganized into currency groups as recommended by the job satisfaction working party. It had been hoped that this reorganization would be immediately followed by the implementation of the on-line computer system but this was unavoidably delayed until June 1976. It was then introduced with very few problems and almost immediately operated at a high level of technical efficiency.

The job satisfaction working party had requested in its report that all jobs be re-evaluated once the reorganization into currency sections had taken place and the on-line computer system had become operational. This job evaluation exercise had to be postponed until January 1977 when the on-line system had settled down. The delay caused morale problems as staff became frustrated at having to wait so long for regrading when the new work structure meant that they were undertaking larger jobs and carrying more responsibility. This frustration was increased by the fact that as staff were struggling to learn to operate the on-line system attempts were made to gain some of the staff savings promised in the report of the job satisfaction working party. Staff who left were not replaced and the exchange department reduced its labour force from 61 to 53 over a period of four months. Although staff had agreed to run the department with fewer clerks this rapid reduction in numbers at a time of major change caused some stress and management quickly responded by increasing the staff complement by two. Nevertheless this did not completely solve the staffing problem. The head of the exchange department was worried that the job satisfaction working party had recommended too large a reduction in staff and he therefore asked for his department to be surveyed as part of the bank's clerical work improvement programme. This showed that rather than attempting to reduce staff numbers further, the department required two additional staff to enable it to function effectively. The staff complement was eventually increased by two in March 1978.

Further problems had, however, arisen at the end of January 1977 when the job evaluation exercise was completed by the bank's job evaluation officers. A number of staff now found that despite the reorganization and the more varied and demanding work, their jobs had not been upgraded. These staff appealed against this decision and as a result some were allotted a higher grade, but this dispute also damaged morale.

9.6.1. The Clerks' Perception of Integration

The author carrried out her post-change job satisfaction survey at the end of March 1978 when the problems of grading and staff numbers had been resolved. Nevertheless the prior existence of these problems almost certainly affected answers to the questionnaire. This contained a number of questions asking for comparisons to be made between the new on-line system and the old batch system which meant that it was completed principally by clerks who could make this comparison. Twenty-six clerks completed the questionnaire, twenty-four of them indicating that they had had experience of the batch system. Answers therefore came from clerks who had worked in the exchange department for at least two years and who might be starting to experience frustration through the scarcity of promotion opportunities within the department. The results of the post-change survey are set out in Table 5.

On most of the facets of work there is little difference between the pre- and post-change results. A larger percentage in the post-change survey say that work is usually or always routine; a smaller percentage say that they have little or no chance to see a piece of work through from start to finish, but a larger percentage say that they would like to have more opportunities to do this. The substitution of an on-line for a batch computer system has not increased the amount of challenge in work, and 45% of clerks say that it has reduced this. It has made no difference to the opportunity to take decisions (71%) but it seems to have somewhat reduced the dependency of one clerk on another and to have decreased pressure and work peaking through a smoother and faster workflow, making deadlines easier to meet. It has also provided clerks with more information (59%) and with a faster feedback on errors via the VDUs (54%).

Table 6 indicates that in the post-change study fewer staff say that they enjoy their work very much, that they are strongly involved in their job and that their present job is close to their ideal.[3]

More people say that they feel very little involved in their job although a smaller percentage report "taking my job as a whole, I do not enjoy it". The majority (82%) say that they "quite enjoy work". In the pre-change study there had been a greater polarization of attitudes in answer to this question, with only 56% saying that they "quite enjoyed work". However, it was clear that despite the reorganization of work and the introduction of the on-line computer system, there had been no major increase in job satisfaction.

These results now have to be explained. The international bank had taken all the steps looked upon with favour in the humanistic literature. It had made an increase in job satisfaction a specific change goal and it had given the bank clerks, through a democratically elected representative group, responsibility for redesigning the organization of work so as to achieve this increase. The results of the post-change attitude survey should, in theory, have shown more satisfaction with work and with the computer system than the pre-change results. Unfortunately the evidence shows that this was not the case.

Table 5. International bank (post-change). Opinion on job structure (in percentages)

Category	Aspect		Statement	FACT/PREF	N = 36	N = 26
VARIETY	Routine and challenge		Work is usually or always routine	FACT	19	62
			There is too little challenge	FACT	48	46
			There is too much routine	FACT	67	69
			Almost everything has rules or set procedures	PREF	31	40
			Can hardly ever choose our work methods	PREF	31	19
DISCRETION	Means choice		Can hardly ever choose work sequence	FACT	17	16
			Can hardly ever choose next task	FACT	19	27
			Would like more choice of work methods	FACT	31	36
			Would like more choice of work sequence	FACT	20	26
			Would like more choice of next task	PREF	30	30
			Can hardly ever make decisions, use judgement	PREF	36	42
	Knowledge choice		Would like more opportunities to make decisions, use judgement	PREF	57	62
			Has very little or little chance to see a piece of work through from start to finish	FACT	42	27
			Would like more chance to do this	PREF	54	70
JOB IDENTITY	Task cycle and integration		Inefficient work by others can almost always or usually, hinder me	FACT	70	69
			Inefficient work by me can almost always or usually hinder others	PREF	83	62
JOB RELATIONS	Work dependency		I would like there to be less work dependency	FACT	31	38
			There is too little or barely enough time to do work	PREF	70	44
GOALS	Deadlines			FACT		

Comparing the old batch with the new on-line system

Statement	%
There is more challenge	5
I can take more decisions	12
Deadlines are easier to meet	25
There is less dependency	39

221

Table 6. Positive and negative feelings about work

Negative feelings about work

	FACT	FACT	FACT	FACT	FACT
	I hardly ever get a sense of achievement	Taking the job as a whole I do not enjoy it	I am very little involved in my job	My present job is far from my ideal job 1/2	Time drags 1/3 of the day or more
Batch system N=36	26	20	6	31	22
On-line system N=26	24	11	22	35	29
	I get this less often 36	I enjoy my job less 25	I enjoy my job less new on-line system		Time drags more often 30

Positive feelings about work

	FACT	PREF	FACT	FACT	FACT	FACT
	I get sense of achievement almost every day	I would like to get sense of achievement more often	Taking the job as a whole I enjoy it very much	I am strongly involved in my job	My present job is close to my ideal job 5/6	Time never seems to drag
Batch system N=36	34	78	23	28	23	50
On-line system N=26	28	83	8	12	8	42
	I get this more frequently —	Compar-ing	I enjoy my job more now 21		old batch with	Time drags less often 5

9.6.2. Management's Perception of Integration

The head of the foreign exchange department believed that the reorganization of work into currency groups had achieved the aim of enriching the job content of the clerks. Within each section each clerk was now able to deal with exchange operations, fixed deposits and loans and with sight and notice accounts, and all clerks could operate the VDUs. This meant that jobs were now more varied and this increased responsibility had led to most of them eventually being upgraded. He now had very few complaints from staff about the nature of the work they had to do. He was less certain that the on-line system had made any major contribution to job satisfaction. Both the batch and the on-line system had caused work to be more bitty and less integrated than the manual system. However, he confirmed the view expressed by a majority of clerks in the post-change survey, that the VDU system enabled errors to be quickly identified and corrected and that queries could be handled more rapidly and efficiently as a result of the on-line system's enquiry facility. The designer of the system made the interesting suggestion that the clerks' ability to identify errors as soon as data was inputted might have reduced job satisfaction and removed some feelings of achievement. Previously a source of job satisfaction had been finding out the reasons for errors and discrepancies and rectifying these. This task had often involved quite complex search processes and some clerical detective work which the on-line system had eliminated.

However, the reason that job satisfaction had not increased to a greater extent seemed related to problems not associatd with either the new organization of work or the computer system. First, in addition to the grading and staffing problems already described there had been a tightening of bank controls and a requirement that all transactions carried out on a particular day had to be processed the same day. Previously deals completed after 3 p.m. were processed the following day. This ruling was a consequence of a number of international banks sustaining heavy losses because dealing staff had not adhered to the procedures laid down by management but it had the effect of lengthening the clerks' working day. Second, the reorganization of work had not solved the department's promotion problems. Once a clerk became a section head at the age of about 25, promotion or an increase in salary could only be achieved through moving to another part of the international bank or to another bank. The two tier recruiting system adopted by most banks in recent years made the problem worse by creating one kind of promotion channel for graduates and a different one for school leavers. Third, the bank had always tried to recruit bright staff and had succeeded in doing this. Now, however, the difficult state of the labour market for young people meant that very intelligent boys and girls wished to join the bank and the routine nature of the work did not fit well with their expectations or ability. In the view of the head of the foreign exchange department the reorganization of work into currency sections had made work more varied but it had not reduced routine. Junior staff were doing more jobs but they were still doing "junior" jobs. The vast amounts of money involved and the strictures of the bank inspectors required that only senior staff signed documents involving the payment of money.

To sum up, the reorganization of work and the on-line computer system fitted well together and appeared to have produced job satisfaction and efficiency gains, although the excellence of the computer system in identifying errors at source reduced the clerks' opportunities for problem solving. There were two factors which prevented work being made more interesting and challenging. The first was the requirement of the bank's inspectors that only senior staff should sanction the payment of money, the second was the decision taken in 1971 to divide up exchange operations and customer services and to allocate the latter to an accounts and payment department. This effectively removed from the staff of the exchange department most of the opportunity for interaction with customers.

9.6.2.1. *The Computer's Contribution to Better Management Information and Control*

Both the head of the foreign exchange department and the systems designer agreed that the on-line system had contributed to improved control through providing the user with better information. Information about customers and about the bank's accounts could be called up on the VDU. As yet it had made little contribution to information for strategic planning but this would be achieved once the international bank transferred to the clearing bank computer in 1980.

9.7. MANAGEMENT'S EVALUATION OF THE SYSTEM

The head of foreign exchange and the systems designer agreed that the original objectives of streamlining operations and providing a faster and more accurate data processing and information system had been achieved. The human objectives of saving staff and improving job satisfaction had been only partially achieved although the reorganization of work had greatly contributed to the latter. In answer to the question ''what aspects of the system best measure up to your idea of a successful computer system'', the systems designer said that it was (1) reliable and (2) met the requirements of both the exchange department and the dealing room. It had the disadvantage of being rather inflexible and difficult to modify easily. The head of the exchange department's opinion was that it was (1) faster, (2) provided a better service to customers and (3) provided a very good enquiry facility. The dealing room was seen as the principal beneficiary of the on-line system. Dealers could now obtain up-to-date information on cash limits and the state of deals at any time of the day.

9.8. CONCLUSIONS ON THE INTERNATIONAL BANK ON-LINE SYSTEM

9.8.1. Goal Setting and Goal Attainment

This was a successful system in that all the technical goals were easily achieved. Human goals were also achieved in the sense that the introduction of a new computer system provided an opportunity for a participative exercise in work design which contributed to improved job satisfaction. A number of problems arose from the fact that staff saving had been made a system objective. Recognizing that the new form of work organization would lead to higher grades and therefore higher salaries, the job satisfaction working party suggested that the greater efficiency resulting from the combination of currency groups and an on-line system would enable the department to operate with a lower staff complement. In consequence staff who left were not replaced and this put the department under a great deal of strain when the on-line system was being introduced and staff were learning to use it. Almost certainly the working party over-estimated the savings that could be made.

9.8.2. Adaptation

Adaptation was greatly assisted by the participative approach to the design of the human part of the system. The clerks themselves analyzed their own problems and chose their own form of work organization. There were no criticisms of the change-over to autonomous currency groups in the second survey. To assist adaptation the head of the foreign exchange department introduced formal training courses in different aspects of foreign exchange activities and these proved very successful. Regrading did cause some problems for despite the increase in work responsibility the bank's job evaluation team did not permit the upgrading of all jobs. This situation was rectified after appeal but it caused a drop in morale in the exchange department.

9.8.3. Integration

The new social and technical systems fitted well together but the two factors of bank controls and fewer errors requiring investigation inhibited the extent to which work could be enriched. Any major improvement of work interest in the exchange department would require the return to it of some of the customer services removed in the 1971 reorganization. The control problem highlighted a conflict of values within the bank. Whereas departmental and senior management were in favour of increasing the responsibility and discretion of the clerks, the bank's control system, represented by the bank's inspectorate, required that certain activities should only be carried out by senior staff.

9.8.4. Pattern Maintenance

The maintenance of a state of equilibrium in the exchange department appears to require the solution of a number of personnel problems. One is the department's absence of promotion opportunities, which is a major cause of high labour turnover. A second is still that of how to provide intelligent staff with enough work interest. Although the new work organization has made an important contribution to this, the tight labour market for young people means that the bank is now able to recruit staff of very high calibre indeed. There is therefore likely to be an increasingly large discrepancy between the job expectations of these young people and the kinds of jobs which the bank can provide. When labour was in short supply this problem was tending to solve itself by compelling the bank to recruit older part-time women clerks who have a higher tolerance of routine work.

10. Conclusions on Systems Design in the Two Firms, the Government Department and the International Bank

Churchman has argued that the relationship between scientists/technologists and managers can take one of three forms: "separate-function" in which there is little interaction between the two groups, the scientist relying on the intrinsic logic of his proposal to induce the manager to implement it; "communication" in which the scientist tries to get the manager to understand why his proposal is of value; "persuasion" in which the scientist tries to arrive at an understanding of the manager's views and interests as a means for overcoming his resistance to change, and "mutual understanding" in which the two come together and try to arrive at a common understanding of each other's needs and interests [Churchman 1965]. The Chemco approach in the two computer systems which have been described would seem to be based on successful attempts to achieve "mutual understanding" at management level, with the technologists responding to user management's desire for greater efficiency and a recognition that the computer could contribute to this. Attempts were also made to achieve a similar mutual understanding at clerical level through the setting up of committees and the development of close informal relationships within the user departments. Where the change was relatively straightforward as in the home sales section this approach worked well. It worked less well in the export sales section where the complicated nature of overseas sales meant that the system was more difficult to computerize. There was also the problem that when both the batch and on-line systems were introduced the department was under stress, the batch system being introduced at the time of an export boom, the on-line system's introduction coinciding with the retirement of experienced clerks. These difficulties meant that the clerks in export sales did not find the new technology easy to adjust to and a good socio-technical integration throughout the department was not easily attained.

Perhaps the most striking thing about the government department case study is how successfully a "mutual understanding" position was achieved. The reason for this success is clear. Without exception the computer technologists who were responsible for the design of the system were themselves previously in the roles of the users whose system they were designing. The Inland Revenue systems designers had all been in the Inland Revenue service and most of them had been district tax inspectors. There was therefore complete understanding of the Civil Service culture and values and considerable understanding of the operational needs and objectives of the user groups. This meant that there were few disagreements over goals and strategies between the systems designers and senior and local management. Goals were clearly defined at the start of the design process and an evaluation was made at the end of how successfully they had been attained. Those goals that were set, were achieved. The goals chosen for the new computer system were strongly influenced by the fundamental Civil Service value of providing an equitable and efficient service to the public and by the Treasury requirement that computer systems should pay for themselves over a ten-year period. The importance of maintaining an excellent public service led to a policy of computerizing existing manual procedures so that manual and ADP systems could fit easily together. The need for systems to be cost-effective led to staff saving objectives.

The Inland Revenue was the only one of the three government department systems studied to include job satisfaction goals in its design objectives. Inland Revenue staff were accustomed to complex problem solving work and would have reacted badly to a computer system which increased work routine, and the systems designers were appreciative of this fact. Strategies to enable staff to adapt easily to new technology were hampered in the early stages of the Inland Revenue implementation by the need to bring staff from a decentralized system of Inland Revenue offices to a new centralized complex, Centre 1, but an appreciation course and pre-change training assisted adaptation to new work procedures and a new technology. The socio-technical integration of the system was generally good, but integration between the internal organization and the external environment was less good with the Centre 1 centralization concept greatly reducing opportunities for face-to-face interaction with the public. At the time this system was designed it was not easy to predict how the introduction of a computer would affect jobs and job satisfaction. The designers of the Inland Revenue system were far-sighted in their recognition of the importance of paying great attention to both the needs of the public and those of clerical staff. The British Civil Service has always tried to be a concerned and humanistic employer and an increased recognition of the importance of job satisfaction is leading today to a strengthening of human values and a keen management desire to introduce change in a manner which is sympathetic to the needs of staff. This value position is clearly shown in the following comments by Keith Robertson, a consultant attached to one of the personnel management divisions of the Civil Service department. Referring to current experiments directed at increasing job satisfaction, he says,

> It is the processes of change, rather than the changes themselves, which matter and which are going to have a lasting effect. Experience must be generated, forms of participation hammered out. In the present series of studies, for example, many hundreds of ideas for change have been developed in "brainstorming" sessions with many different groups . . . . There is an enormous amount to be done, and done urgently, if the Civil Service is to tackle its most pressing problem of the under-utilization of talent at all but the most senior levels; if jobs are to be brought more into line with aspirations, so that good young staff can be recruited, retained, encouraged and inspired to rebuild in a new environment and on a larger scale the very best traditions of public service. (1973)

In Asbestos Ltd. and the international bank the enhancement of an existing batch computer system to on-line was perceived as an opportunity for achieving important personnel gains. The strategy adopted by these firms was not, however, Churchman's mutual understanding, but a more revolutionary one of "self-determination" in which the clerks themselves became the creators of their own work organization. In Asbestos Ltd. the values behind this strategy were shaped by a recognition on the part of the systems designers that they had not always associated previous computer systems with clear human gains, and an association with the Manchester Business School which gave them a knowledge of the techniques that could be used to analyze and design human systems. In the international bank management was anxious to reduce a high rate of labour turnover in the foreign exchange department involving a loss of experienced clerks the bank could not easily replace. The systems designers were also aware of the complexity of foreign exchange operations and believed that the group with expert knowledge, in this instance the clerks themselves, was the one best fitted to design the organization of work.

In both of these situations this giving of design responsibility to the clerks had three clear gains. First, it enabled the clerks to develop forms of work organization that fitted well with their job satisfaction and efficiency needs. Second, the processes of analysis and design provided a learning opportunity for the clerks with the result that they were able to gain a detailed understanding of the functions and problems of their department. They were also able to gain an excellent appreciation of how the computer could be used to assist with these problems. The result in both situations was a good integration of the social and technical systems in which the use of the computer and the organization of work reinforced each other in providing efficiency and job satisfaction. This good internal integration had an impact on the external environment by enabling an improved service to be given to customers. Third, there was the psychological gain that the clerks were committed to the new computer systems and to the reorganization of work because they had played a major role in creating these.[1]

There are two potential problems with this "self-determination" approach. First it tends to exclude departmental management from the design activities and this could result in management later refusing to accept a solution put forward by the clerks. Therefore management must have the option of joining the design group if they wish to, even though in some situations this may inhibit the group's freedom of thought. If they decide not to join then careful communication lines

---

1. Unfortunately, in the international bank these integrative gains were offset by a number of personnel problems. A lack of promotion prospects in the exchange department meant that clerks continued to be lost to other departments and banks. Also, because of the dramatic change in the labour market after 1975 new recruits tended to be exceptionally well-qualified and of high intelligence with the result that the more varied work of the currency groups still seemed routine. More extensive job enrichment was prevented by the bank's control requirements and by an earlier reorganization in which customer service activities were transferred to another department.

must be developed and maintained to ensure that management is kept completely informed of what is taking place. The provision of a steering group can overcome this difficulty, as departmental and senior management can sit on this [Mumford, Land, Hawgood 1978]. A second problem is that although the "representative" participation approach adopted by the international bank means that the clerks do design their own work organization, only some of the clerks are involved in this and there is the question of whether a representative group really represents. The author has tried to overcome this problem in other firms, for example, Rolls Royce, by using an approach she calls "consensus" participation. The design group now feeds its ideas back regularly to its colleagues in the department and the final choice of work organization is made by the department as a whole [Mumford, Henshall 1978]. This approach can be said to be truly democratic but experience has shown that many conflicts of interest within the ranks of the clerks have to be resolved before a system evolves that fits everyone's needs.

Another interesting feature of the "self-deterministic" strategy is that the systems designers have also to learn new skills and a new role. If they are initiating socio-technical design then they too must understand how to carry out this kind of design. This requires a good knowledge of theories of job satisfaction and of work design principles, particularly the socio-technical approach to work design. In both Asbestos Ltd. and the international bank the author acted as consultant, but if these two organizations continue with a participative approach they will want to do this on their own. The systems designers will then have to assume the roles of teacher, adviser and consultant and abandon their traditional role of the expert who is sole designer.

Finally the "self-deterministic" approach requires a particular set of values from all the participants. The systems designers must believe in the importance and viability of designing the human part of what have in the past been regarded as purely technical systems, and be willing to let the user assume responsibility for this area of systems design. Management must believe in participation and be willing to hand over an important design task to their subordinates, and the clerks themselves must be prepared to take on the design task and not see design as something for which only management or experts should have responsibility. Both Asbestos Ltd. and the international bank had systems designers, managers and clerks who shared important values on participation and the organization of work. These groups were therefore able to cooperate effectively in the design of their new computer systems and to gain the advantages of the user interest and motivation which accompanies such cooperation.

11. Values and the Change Process

11.1 THE MODEL

The research data described in the last four chapters now needs to be reviewed and commented upon so that it can be used as a basis for contributing to "theory", by which is meant an understanding of the interrelationships between variables in the change situations, and "practice", the provision of a set of analytical tools or normative proposals which will facilitate the management of change by *all* groups involved in a change process. Figure 1 sets out the author's perception of some of the important variables associated with change or "changing" [Bennis 1966: 100]. It should be noted that this form of diagramatic presentation tends to make processes which are inherently complex and disordered look as if they are neat, sequential and logical and it is important to stress that the author does not see change in this way.

The model covers what have been seen as the two main stages of a change process: (*a*) the creation of an idea and its development and (*b*) the introduction and adoption of an idea [Knight 1967]. It starts with the decision to make a change, although it is recognized that complex processes can precede this point in time and these have been studied by the author and documented elsewhere [Mumford and Pettigrew 1975]. All the aspects of change shown will be influenced by, and will influence, the values of different groups associated with the change, and it is this value impact that this book is most concerned with.

There are a number of intellectual positions on what planning for change should encompass, starting with the now largely discredited "rational planning" approach in which experts set goals, search for ways of achieving these goals, evaluate the consequences of each alternative strategy and choose one on the basis of its optimality in attaining the goals [Dale 1962; March and Simon 1958].

Other approaches include "adaptive" planning which recognizes the importance of timing and uncertainty in the change process and involves environmental monitoring and response to feedback [Hekimian and Mintzberg 1968; Beer 1969], and "contingency" planning which is concerned with events which may occur in the future and requires the formulation of a number of alternative plans, the final choice depending upon the eventual situation. In addition there is "experimental" planning, formulated by Sackman [1971: 534] in which the planning process is seen as a set of hypotheses which have to be clearly formulated and tested out as an aid to learning. And there is the approach which is beginning to

Figure 1. Values and the change process

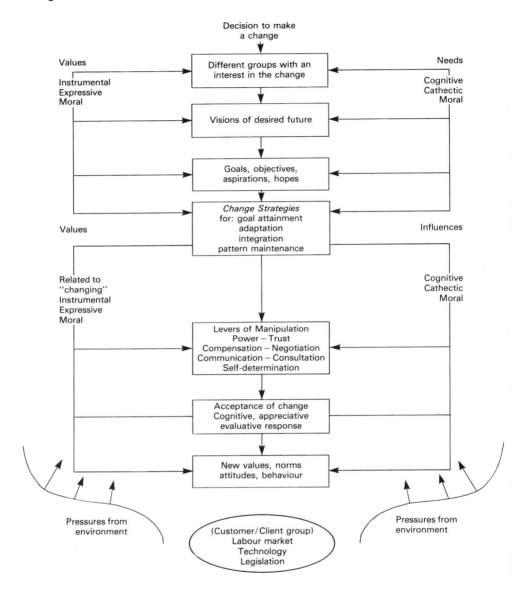

find favour at the present time of "cooperative" planning which has as one of its objectives the recognition and negotiation of different interests [Hekimian and Mintzberg 1968; Mumford 1976]. This approach sees planning as a social rather than an intellectual process, but with intellectual elements contained within it [Bauer 1966]. Lastly, there is the theory of Weick [1969: 101] who sees planning as thinking in the future perfect tense. In his view a plan works not because it accurately anticipates future contingencies, but because it can be referred back to analogous actions in the past. Planning is therefore of no use in novel situations.

Planning for change is also carried out with different ends in view. Traditionally an important end has been an improvement of efficiency and although this end still forms an important part of most change programmes, other goals are increasingly being associated with it. Vickers perceives a need to develop "appreciative" systems or systems which incorporate enhanced physical and social environments for all their members [Vickers 1968]. The socio-technical school of thought believes that social systems and technology must be tightly integrated so that they reinforce each other and create both efficiency and the satisfaction of human needs [Trist and Bamforth 1951; Herbst 1974].

The author's philosophy is that change processes need to be adaptive and cooperative to cater for changing needs and different interests. If possible they should also be experimental so that the objective monitoring of cause and effect can lead to a cumulative build-up of knowledge. Planning is therefore creating a structure in which these things can happen, rather than formulating precise steps towards a goal. The model in Figure 1 reflects this value position.

Before the research data is related to the model, the model itself requires some explanation and clarification. The three top boxes depict different groups with an interest in a proposed change, seeing or wishing to see the change in terms of visions of a desired future, and translating these visions into a set of goals, objectives, aspirations and hopes, depending upon the nature of their role in the change process. Interested groups when computer systems are being introduced will be the technical systems designers, the managerial and non-managerial users, both of these groups being direct recipients of the effects of the system; indirect recipients of these effects located within the organization such as functions or departments which interact with the users, and indirect recipients outside the organization such as clients or customers. All of these groups have an interest in the kinds of goals and objectives that are set, the way these are attained and the change outcomes. Their intervention in the change process and response to it will be influenced by both their values and their interests. Values being defined as "a concept of the desirable which influences choice and behaviour" [Kluckhohn] and interests as a "desire for special advantages either for oneself or for groups with which one is identified" [Neal 1965: 9]. Interests are more likely to be short-term and superficial and values to be long-term and deep.

Values and interests, particularly those associated with one's occupational role, lead to different ways of perceiving the world [Mason 1969] and all parties involved with change are likely to be as much influenced by emotional and social factors as they are by objective and rational factors. Mumford and Pettigrew

[1975: chap. 5] have already made a number of detailed studies showing the influence of organizational politics on strategic decision taking. It must be recognized that the interests of people and of organizations are multiple and complex and that when faced with a new situation full of ambiguity some groups may be unsure as to where their interests lie. This can result in shifting attitudes until the sphere of their self-interest becomes clear [Bauer 1966]. In contrast, the values of the technologists may lead them to have an extremely clear vision of their role and objectives. This can have its advantages for values are the basis of commitment and it may be functional for organizations to have a hard core of people who are dedicated to their mission [Katz and Georgopoulos 1971]. It may also have the disadvantage of causing what has sometimes been called "structured perspective" or "tunnel vision"; a situation in which the interests of other groups are not understood or receive little sympathy [Bjorn Andersen and Hedberg 1977: 127]. A recognition of this danger is leading to the present-day belief that groups with interests in a proposed change should be able to represent themselves through participating in the design processes [West and Churchman 1968; Mumford, Land, Hawgood 1978] and that successful change is a result of cooperating coalitions rather than the efforts of a single expert group [Bauer 1966].

Visions of a desired future will also be conditioned by values, interests and roles. It has been argued that the desired future of many computer technologists was, and perhaps still is, one that embodied an increasingly sophisticated use of machines with the excitement of designing "frontier of knowledge" systems [Martin and Norman 1970]. As Sackman has said, referring to the 1950s and 1960s,

> Computer professionals lived in an esoteric world of their own that became increasingly isolated from the human and social needs they were serving. And why not? Demand for computer services continued to rise exponentially, salaries kept going up, sheer experience reigned supreme, and the vicious cycle became self-sustaining. This is the Dorien Gray portrait of the other, the hidden, side of the computer coin. [1971: 50]

Sackman also argued that this response was logical because computers were originally created to count and measure. Therefore the professionals who first developed and used computers were interested in abstract ideas and their embodiment in hardware, rather than in human behaviour in using such devices. The requirements of the technology therefore influenced recruitment and this led to the dominance of certain values which endured over time. Today, the development of distributed computing, which is much more user oriented, requires technologists who will work closely with users; the technical and human needs of tomorrow therefore require different visions of the future and these visions may fit the values of a type of person who is not the same as the technologist of the past.[1]

---

1. There is some evidence that technologists who have developed large centralized systems are not interested in developing distributed systems [Personal communication from Director of NCC).

The manager is likely to have a vision of the future that incorporates both organizational and personal goals. Organizational goals in most situations will be related to improving efficiency, but there may also be sub-goals such as establishing good human relationships and an organizational ability to adapt easily to change. Realizing organizational goals may contribute to personal goals for the manager may believe that if he improves the efficiency of his department he will be rewarded with promotion or an increase in salary. Most managers will therefore focus on futures that meet both organizational and personal goals. The clerk in contrast may have a less clear vision of the future. If he likes his existing situation, his preferred future may be one in which it does not change. Or a desirable future may be associated with an increase in pay, or an increase in job satisfaction. Almost certainly, if he is not a participant in the systems design process, he will be unsure about what the future holds in store once the new system is implemented, and is likely to feel anxiety about this. The customer or client's interest is in service and he may be unaware that a new computer system is being introduced until the service he receives either improves or deteriorates, or he is asked to accept a new set of procedures. If his first experience of the change suggests that his future relationships with the firm will be unsatisfactory, then he has the option to severe these and take his custom elsewhere or if he is a client of a public service, to complain to his M.P.

But, the visions of all these groups are likely to be over-simplifications. For what has been called the "cognitive limits on rationality", the fact that an individual can only take in so many things at one time, will cause as yet unknown situations to be defined in partial rather than total terms [March and Simon 1966: 151; Miller 1960; Osgood 1964]. The technical vision of the systems designer may not include the fact that people will not use his system because they see it as in conflict with their interests. The efficiency vision of the manager may not include the stress of unreliable hardware and a system which is "down" for long periods of time. The clerk's vision may be of an immutable and deterministic technology to which he has to adjust because it cannot adjust to him.

Simplified visions lead to simplified goals. The systems designer who wishes to optimise his technology will set only technical goals and be delighted when he finds that he has achieved these. The fact that problems have arisen because certain goals have not been set will be of no concern of his and he can dismiss these as the deficiencies of a user who does not have the intelligence to operate his excellent system [Bjorn Andersen and Hedberg 1977]. Similarly the manager who interprets efficiency as an increased speed of processing and lower costs through a reduction in his staff can find his vision of the future has become reality and may be unconcerned that he now runs an unhappy rather than a happy department. He has achieved those things on which he will be judged by his superiors. The clerks will have aspirations and hopes rather than clearly specified goals but they are likely to find that because no-one knew what these were, and nobody cared very much, if these do become reality it is through chance and not intention.

The design of computer systems, at least as described in retrospect and this may be a distortion of the truth [Weick 1969: 102], appears to resemble, or aim at, what has been called an algorithmic or "Cook's tour" approach in which the change process takes the form of a series of steps forward en route from goal setting to the final design of the system [Steiner 1969: 7; Ackoff and Rivett 1963: 34]. It is less likely to take a "Lewis and Clark" approach in which decisions are taken as each problem is encountered, with the manager and the systems designer working closely together to spot the next problem ahead.[2] Another name for this approach is "incrementalism" [Popper 1963: 158; Etzioni 1968: 268]. Therefore, systems design and implementation are not adaptive in concept; if strategies are changed then this change is made reluctantly and is seen as a deviation from the "plan".

The four following boxes in the model (Figure 1) are "change strategies", "levers of manipulation", "attitudes to change" and the development of "new values, norms, attitudes and behaviour". Strategies have been defined as operationalized theories of the world; coherent sets of rules which guide an organization's search for problems and solutions [Hedberg and Jonsson 1976: 2] but they are more than this. Ozbekhan describes them as essentially to do with people.

> People who receive instructions or give them, people who meet in councils and committees; people who are called upon to make decisions in support of which little information is usually available; people who are baffled by their responsibilities and often uncertain as to both their private futures and the future of the function they are carrying out. [Ozbekhan 1969: 136]

Weick takes this idea a stage further by seeing control, which in the author's view incorporates the notion of strategy, as being accomplished by relationships, not people. The way processes unfold and their consequences for other processes is determined by the specific, causal relationships which exist. People are the medium through which these relationships are made actual; but it is the relationships not the people that exert the influence [Weick 1969: 37].

In this book the framework for examining the change strategies and processes associated with the design and introduction of computer systems has been derived from Talcott Parsons and covers the attainment of goals, securing adaptation to the new system, the integration of the various sub-systems associated with the change and the maintenance of the new situation in a state of equilibrium so that it avoids moving into what Beer has described as a "crisis" or a "suicide" mode [1972: 294]. One way of understanding the values that lie behind the change processes in an organization is through an examination of the kinds of goals that were set, who was involved in the decision to set particular goals and the kinds of actions that followed the goal setting process. Weick makes the point that action is not necessarily related to goals [1969: 37] but the researcher who is not part of a change situation has no opportunity for observing action at first hand, whereas he

---

2. It will be remembered that Lewis and Clark were two explorers who in 1805, at the behest of President Jefferson, opened up the Pacific coast of the United States; an area until then inhabited solely by Indians.

or she can usually obtain information on what people were trying to achieve at the start of a change process and the extent to which they believed they succeeded in their aim. The author has pointed out elsewhere that there is little information on how members of systems design teams perceive their tasks and roles and an examination of goals that were set, or thought to be set, and the strategies used to attain these can throw light on this role definition [Mumford 1966, 1972, 1975]. Goals that meet the needs and interests of all the various groups affected by a proposed change can be said to be desirable, although the opposite has been argued [Weick 1969: 37], but they are rarely attainable. Most futures are negotiated, with each group making some trade-offs to meet the needs or demands of other groups [Touraine 1974: 172] and the nature of the trade-offs will depend on the power relationships in the situation [Crozier 1964: 298; Dalton 1959: 263]. Previous research carried out by the author and a colleague has investigated the nature of these power relationships; it was not possible to do so in the present project [Pettigrew 1973; Mumford and Pettigrew 1975].

Adaptation requires flexibility and stability for there must be movement from one system state or set of relationships to another without too traumatic a disruption [Campbell 1961: 101-142; Hollander and Willis 1967]. Weick points out that in some situations these two things may mutually exclusive [ibid: 39] but nevertheless it is what most organizations try to achieve and they do this through the use of various levers of manipulation. For example, the authoritarian command legitimated through the holding of a power position; providing compensation for the acceptance of change or "buying" innovation in; negotiation so that different interests are catered for in the change; communication to provide understanding; consultation, which implies more than one group having a say in the decisions that are taken, and self-determination, or letting the group with the major interest in the change take control of it. In addition "trust" has been included although it is not usually seen as a manipulative technique. Nevertheless, the fact that one group trusts another to look after its interests and therefore does not question goals or strategies, can be an important, if rare, facilitator of change. These levers are not mutually exclusive, some or all may be used at one and the same time and the choice of levers will be influenced by the values of those responsible for system implementation.

Integration is likely to be more difficult to achieve than adaptation and successful adaptation does not necessarily imply that integration has taken place. Integration is seen as an aid to system viability although it has been argued that too tight integration can be organizationally dysfunctional in that it inhibits change [Gowler 1972: 25]. The case studies have attempted to show the extent to which the clerks in each change situation experienced role integration, with a good fit between their job needs and expectations and the requirements for their occupational role; they have also illustrated the impact of technical change on sociotechnical integration or the successful interaction between social and technical systems, and on environmental integration, or the establishment of successful relationships between internal and external organizational environments.

Pattern maintenance is more difficult to evaluate as it requires the researcher to observe an organization over a long period of time, which has not been possible in this research. But predictions can be made about the likely pattern of events if conditions in the environment do not change radically from the situation that existed when the research was carried out.

The acceptance of change requires, in Parsons' terms, cognitive, appreciative and evaluative responses from all the groups concerned. Systems designers, senior and user management, the clerks and the customers must all understand the nature and logic of the change; must approve of and like the results and must regard the decision to change as a correct one in terms of their own interests. Lastly, the establishment and successful operation of a new system requires and will stimulate new values, norms, attitudes and behaviour. If the system fits with cognitive, cathectic and moral needs, then these values and attitudes are likely to assist this successful operation. If the system does not fit these needs then the new values, norms, etc., may be directed at destroying the system or ensuring that it works ineffectively.

The influence' of environment on all of these aspects of change needs to be stressed. Environment is usually seen as exerting pressures on an organization; it is less often described as something which the organization itself tries to influence and control although some researchers have made this point [Perrow 1972: 199; Randall 1973]. Yet it is clear that both of these things happen. The author studied over a number of years the design of a very large computer system in a mail order company in which the system was fundamentally rethought twice as new equipment appeared on the market, presenting the systems design team with new technical opportunities and possibilities. In the government department, the system was designed so as to disturb the external client environment as little as possible and few members of the public were aware that a computer had been introduced.

All aspects of the change process, the interaction between one internal group and another and between the internal and external environments create a complex and very fluid set of activities in which new actions are constantly stimulated and responded to, this process in turn setting off a train of new actions and responses. At this point the author will redraw the formal, and misleading, diagram in Figure 1 and replace it with the untidy diagram in Figure 2, which more closely resembles her vision of the change process. The dynamic nature of these relationships generates both uncertainty and in most groups, a desire for control [Lawrence and Lorsch 1967: 27; Argyris 1972: 73]. Beer and Hedberg [1974, 1975] see this control or management of uncertainty as the responsibility of what they call a meta-system. In the context of systems design a meta-system is a higher order system which provides the conceptual structure which holds lower order sub-systems together. An important part of this structure will be values. For example, top management acting as a meta-system, could insist that design processes are carried out using democratic design principles. The author's own approach has been to set up steering committees composed of senior managers and union officials which provide guidance to design groups on the extent to which design alternatives conform with or deviate from the firm's policies and values

238

Figure 2. Values and the change process

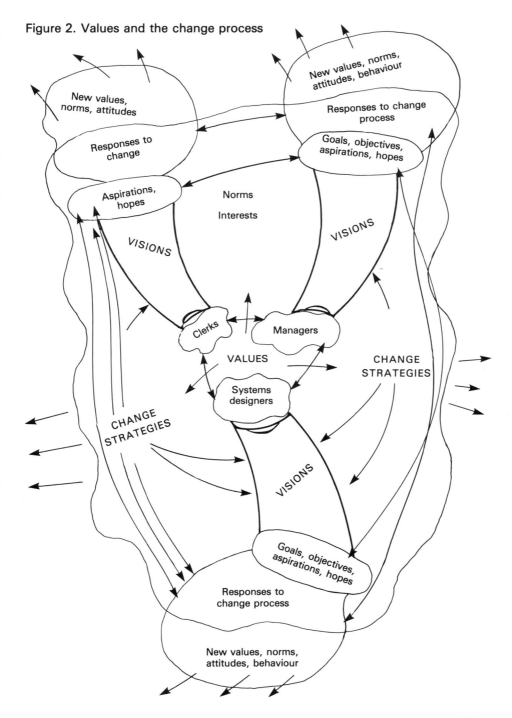

New values, norms, attitudes

Responses to change

Aspirations, hopes

VISIONS

New values, norms, attitudes, behaviour

Responses to change process

Goals, objectives, aspirations, hopes

VISIONS

Norms

Interests

Clerks

Managers

VALUES

Systems designers

CHANGE STRATEGIES

CHANGE STRATEGIES

VISIONS

Goals, objectives, aspirations, hopes

Responses to change process

New values, norms, attitudes, behaviour

239

[Mumford 1977: 87]. Ashby's law of requisite variety postulates that if uncertainty is to be controlled then the controlling mechanism must be of at least as great a variety as the situation to be controlled [Ashby 1952]. This means that any top management group or steering committee that oversees the design of a new computer system must be (a) aware of the complexity of the process they are managing, and (b) have the necessary resources to introduce desired inputs into the change process and deflect undesired stimuli. They should also have the vision to put larger interests, even those related to society, before sectional interests [Barnard 1938: 215]. Vickers lists five levels of control: control by release, control by rule, control by purpose, control by norm and control by self-determination [1973: 182]. He favours the last, describing it as

> a process both individual and social which depends on ethical debate and reflection about changing norms and values and on the policy making which both expresses and generates that debate.

Lastly, it is worth repeating a point that has already been made about visions of the future. Uncertainty is related to perception and perception is related to values. A situation that appears completely straightforward and simple to one person, will be seen as unbearably complicated by another [Duncan 1972]. Most of us cope with uncertainty by a process of perceptual selection which acts as a variety reducer [Knight 1967; Beer 1972: 291] and enables us to work within our cognitive limits. In doing so we provide ourselves with mental security but we may generate major uncertainty for other individuals or groups whose problems lie outside our range of vision.

11.2. VALUES, INTERESTS, NORMS AND ATTITUDES

Values can be regarded as an important gatekeeper which acts as a variety reducer guiding our judgement and choice and enabling us to cope with uncertainty, complexity and ambiguity. Values are a central aspect of all human belief and action; they are integrated into patterns of living in such a way as to be both goals and instrumental tools, and the concepts with which we think are shaped by values [Peattie 1965]. Like the frog whose physiological structure makes him see only objects which he can eat, we focus on those things which fit into our cognitive structures and ignore or pay less attention to those which do not. In the work situation values can be defined as constellations of attitudes and opinions with which an individual evaluates his job and work environment [Pennings 1970]. Values are also the stimuli for change. They specify the conditions under which members of a group should express dissatisfaction and prepare to undertake change [Smelser 1959: 16]. A number of writers have suggested that values imply action and that values which do not have action associated with them are very weak [Etzioni 1968: 12; Kluckhohn 1951: 400].

One of the features of modern society is the differentiation of values that is accompanying the differentiation of life styles. Whereas members of a community, fifty years ago, would broadly accept the same set of values, a visitor to, say, Los

Angeles today would find a wide variety of groups living side by side, from the drug-attracted Brotherhood of Eternal Love to the Children of Jesus or the Mormons [Rittel and Webber 1973]. A similar, although less extreme differentiation of values is occurring in industry and commerce where the pervasive idea of efficiency as the most important goal, derived from the 18th century physics and classical economies, is being replaced by other more humanistic values and goals. With this modification of goals there has been a shift in management philosophy so that the dynamic, forceful, follow-the-leader type of manager is being replaced by one that is more reflective and evaluative [Churchman 1968].

Values are different from interests in that they are deeper and more enduring and the function of values is interpreted in different ways by social scientists and philosophers. Parsons, whose theoretical orientation is to integration and an understanding of what welds groups together and creates stability, perceives values as assisting this integration by setting limits on the kinds of interests that can be expressed. He says:

> It can be taken as a fundamental proposition of social science that no system of the ''play of interests'' can be stable unless the pursuit of these interests is carried out within an institutionalized normative system – a common framework of values, of generalized norms and of the structuring of interests themselves [Parsons 1961].

Other writers such as Marx [1862], Touraine [1974], Dahrendorf [1959] and Gluckman [1955] interpret behaviour in terms of conflict rather than integration and see the interests of specific groups as generating values or manipulating existing values. Almost certainly both arguments are true with some values contributing to the internal integration of the group or society and other values emphasizing the differences between it and other groups.

Norms are the ways in which social values are expressed [Sherif 1966: 113] and generally relate to standards of behaviour. Values are general and explicit, norms are specific and tacit [Vickers 1973: 175]. Vickers suggests that while values clearly produce norms, norms can also influence and generate values. Powerful appeals by politicians or religious leaders can alter tacit standards of what is acceptable and unacceptable and produce a new and enduring set of values. Vickers quotes Wilberforce and slavery as an example, the Womens Liberation Movement might be another example today. He also suggests that a person who makes policy has a right and duty to advocate his norms and values. He should be

> an artist in the creation of coherent and viable behaviour and, like any other artist, he must believe in the goodness, as well as the coherence and viability of the design which he is trying to realise. And even beyond this, he is an artist in shaping the norms and values from which his policy is made. [1973: 178]

The norms of occupational groups take the form of similar standards of consumption, similar family modes, similar standards of dress and decorum [Caplow 1964: 124]. These have the function of giving a group associated with an occupa-

241

tional role an external and internal identity which separates it off from other groups. In the past this deliberate emphasis on "separateness", if associated with a prestigious role, resulted in respect. Today, it is more likely to be received with attack and censure as society pursues its quest for equity rather than hierarchy [Rittel 1973: 155]. Social change frequently affects the status of different occupational roles. Parsons suggests that when change is perceived as leading to a fall in status, groups try to defend themselves by preventing the change taking place [Parsons 1964: 244]. If they are successful in doing this, then they are likely to move into a state of malintegration and tension which prevents the stable institutionalization of the new patterns.

Attitudes differ from values in that attitude refers to an organization of several beliefs around a specific object or situation [Rokeach 1973: 18], whereas a value refers to a single belief. Values and norms occupy a more central position within an individual's personality and cognitive system and are linked to motivation. Attitudes are predispositions to evaluate objects in a consistent and characteristic way. They have also been defined as a predisposition to respond in a particular way to an external stimulus [Allport 1935]. Nevertheless, like norms, attitudes are expressions of particular values.

11.3. VALUES AND ROLES

The focus of interest of this book is on how values influence the attitudes and behaviour of the principal groups associated with a particular kind of change, the introduction of computer systems. Throughout the discussion a relationship between values and roles has been assumed. For a sociologist roles are patterns of expected behaviour with certain effects, for a philosopher interested in social ethics they are clusters of rights and duties [Downie 1971: 121]. An important difference between these two definitions is that whereas social ethics assume that the person who holds the role will know what these rights and duties are, the sociologist does not necessarily assume that a person in a role will be aware of the behaviour expected of him by others. Because of this people in new roles, or old roles in new situations, may experience a great deal of role ambiguity [Kahn et al. 1964: chap. 5]. If an individual or group in a change agent role is continually faced with groups who have unclear or widely differing expectations of him, and if, at the same time, his role is itself dynamic and subject to redefinition, then he is likely to increasingly fall back on his values as a guide to action. Recognizing that action is related to values some professional associations are developing codes of ethics to assist their members to formulate values. This is true of the British Association of University Teachers, a number of associations of professional engineers and the British Computer Society. For example, the Code of Ethics of the Association of Professional Engineers of Ontario states that the engineer "cannot rightly delegate his responsibilities. Therefore he should not accept without thinking standards of values and behaviour suggested to him".

The role expectations that one group has of another can be formed in a number of different ways but are likely to be greatly influenced by observed acts and behaviour with these observations leading to moral judgements [Katz 1969: 1375]. Where groups have only occasional contact and one group is seen as threatening by another, stereotyping may occur [Sykes 1964: 151] and this is used to rationalize or justify prejudice [Allport 1954: 187]. It has been suggested that increased contact and demonstrations that a group's intention is benevolent rather than hostile can assist the elimination of unrealistic stereotypes [Saenger and Flowerman 1954: 237]. However, a group which sees itself as benevolent can still produce consequences which are against the interests of another group, and this has been true of many computer systems in the past.

11.4. SYSTEMS DESIGNERS, MANAGERS AND CLERKS

George Bernard Shaw has said that "every profession is a conspiracy against the laity"; Veblen coined the term "professional psychosis" to describe the narrow view of the expert. Although computer technologists are not yet professionals in the sense that they are required to have specialized qualifications [Prandy 1965: 44; Gerstl and Hutton 1966: 4] they have some of the characteristics associated with professional groups; for example, shared values, a particular approach to problems and a strong sense of group identity [Kuhn 1962: 186; Ben-David 1971: 4; Rittel and Webber 1973: 155]. They have also in the past been able to create dependency relationships in others because they have had access to knowledge from which others were excluded and this knowledge gave them power [Blau 1964: chap. 9; Emerson 1962]. Some researchers have suggested that in an effort to increase this power experts try to manipulate the uncertainty surrounding their expertise [Pettigrew 1973: 234]. The use of jargon is one way of doing this, as is ensuring that knowledge is not readily transferred to another group [Ward 1973: 35]. But today the situation is changing, the credibility of the expert is low and he is increasingly being asked by society to examine both his concepts and his goals [Hitch 1960: 19]. Kuhn, referring to scientists, suggests that once existing approaches cease to solve problems then the expert looks for a new set of theories and may seek these in disciplines far removed from his own [1962: 77]. Today the unsatisfactory human consequences of some computer systems may be influencing systems designers to take an interest in the ideas of social scientists [Mumford 1966, 1968, 1969, 1970, 1972, 1973].

All planned change requires some collaboration between the change agent and the client [Bennis 1966: 105] or, in the context of computer systems, the systems designer and the user manager and his subordinates. Unfortunately these three groups do not usually share the same values and goals. User managers have responsibility for running a department efficiently, achieving targets and meeting deadlines, and a major systems change may temporarily jeopardize these essential activities on which their performance is evaluated [Mumford and Ward 1966]. The user rank and file will be interested in preserving those things they regard as important in the work situation and these will include salary, status and the right to certain tasks and responsibilities, which Robertson has called "job ownership" [1973]. The introduction of a new computer system is likely to threaten some of these. Parsons suggests that if a system is to achieve stability then the values and orientations of the various groups involved with change must be compatible enough to be able to "mesh" with each other so that successful group interaction can take place [Parsons 1964: 52]. This implies more than agreement on what is to be done, it requires a set of reciprocal relationships that act as a stimulus to action. In the next section we will analyze the value differences and similarities in the four organizations described in this book. This analysis will be used to assist an understanding of how these values affected the subsequent change processes in each organization. Because the author considers that each change situation is unique and that the kinds of relationships found in one situation will not be replicated in another, generalizations about the consequences of change will not be made.

11.5. VALUES AND CHANGE IN THE FOUR ORGANIZATIONS

The four organizations described in this book had different approaches to the introduction of computer systems but held similar humanistic values. Three of them, the government department, Asbestos Ltd. and the international bank, set precise human goals related to job satisfaction at the start of the design process. Two, Asbestos Ltd. and the international bank used a participative approach in which clerks played a major role in the development of a new form of work organization into which the computer system would be embedded. The fourth firm, Chemco, did believe in participation and had a considerable interest in job satisfaction, although with the computer system discussed here it was less precise and systematic in the way it tried to achieve these than the other three organizations. If the hypothesis that "values influence the way in which change is carried out and received" is correct then there should be similar sets of values and value relationships in these four organizations.

11.5.1. Organizational Values

Kluckhohn [1951: 408] points out that "real" values can be discerned by careful analysis of selection made in choice situations and that "one dare not assume that verbal behaviour tells the observer less about the 'true' values than other kinds of action" [ibid: 406]. Bauer [1959: 166] suggests that "the outlook of any group upon the world and experiences is determined and reflected to an important extent by the clichés they continually use, by the habitual premises which they accept". These comments provide some support for regarding the value scales as evidence.

In order to reduce a large amount of data the 7-point scale has been split into three categories, X for points 1, 2 and 3 on the scale which depict what has been called a "mechanistic" or theory X-type organization with tight structures and controls [Burns and Stalker 1961: 119; McGregor 1960: chap. 3]. Points 5, 6 and 7 representing an "organic", theory Y, style and a high degree of self-management have been called Y, and the centre point of the scale is designated C. The number of modes falling in each of these X, Y and C categories for each group of clerks, managers and systems designers in each organization has been summed up in Table 1. The data covers answers to the three organizational value questions:

(1) Please tick where you *would like* this firm/department to fit on the (value) scale below.
(2) Please indicate on the scale below what you believe to be the best form of *departmental* structure for clerical staff in general.
(3) Please indicate on the scale below how you see the *general characteristics* of the majority of staff in this department.

These three questions require twenty responses.

In the government department the values of the managers and systems designers were in the Y category (13 out of 20 answers to value questions) with the values of the clerks in the C or Y category. In Asbestos Ltd., the values of the systems designers fell mainly into the Y category (17 out of 20 answers) and the values of the clerks into the C or Y categories, but management has more responses in X. This conforms with the analysis of the firm's change process in chap. 7 where it was shown that the initiative for using a participative approach and making job satisfaction a system goal came from the systems designers, not from management. The international bank's management is predominantly Y (15 out of 20 answers), the clerks C and the systems designers split between X and Y. This fits with the analysis in chap. 9 where it was shown that the stimulus for participation and job satisfaction goals came from management. The split in values is due to the fact that two systems designers were interviewed; one who designed the early batch system, the second who designed the on-line system. The X responses are mainly from the designer of the batch system who was not involved in the later system. In Chemco, where the systems were designed without incorporating job satisfaction goals, management was in the Y category, the clerks in the C and Y categories, but 10 of the systems designers' ansers to the 20 value questions were in the X category. The absence of job satisfaction objectives suggests that management values did not sufficiently influence the systems design process and the X oriented systems designers although attempts were made to use a participative approach.

Figure 3 shows the strength and extent of values in each organization. It can be seen that the values of the clerks are weaker than those of the managers and systems designers with more C responses.

A second hypothesis in this book is that a good fit between the values of all the groups associated with a major change will lead to change consequences which are approved of. This hypothesis is generally supported by the data. Job satisfaction did not increase in the international bank after the introduction of the real-time system but this was a consequence of staffing and grading problems and not related to the computer or the reorganization of work.

In these four organizations there is some evidence to suggest that there was a relationship between values and value consensus and the design, consequences and reception of their computer systems. However, it must not be assumed that

Table 1. Number of modes falling into the X, Y and C categories for the questions on
organizational values
Number of questions: 20

	Government department (Inland Revenue)			Asbestos Ltd.			Chemco			International bank		
	X	C	Y	X	C	Y	X	C	Y	X	C	Y
Clerks	2	11	8 (19)*	1	12	7 (19)	1	11	8 (19)	6	12	3 (15)*
Managers	8	2	14 (16)*	11	2	7 (9)	3	2	15 (17)	3	1	16 (17)
Systems designers	9	1	13 (14)*	8	—	17 (17)*	10	4	6 (10)	14	—	13 (13)*

* Where figures add up to more than 20 there has been one or more double modes.
The figures in parentheses are C (the central point) plus Y.

Note:
The two government departments which requested that their experience should not be included in this book did not make job satisfaction a system objective or use a participative design approach. Their profiles are quite different as the figures below show. The answers of the managers and systems designers to the twenty questions on values fell predominantly into the X category.

	Department A			Department B		
	X	C	Y	X	C	Y
Clerks	3	7	10 (17)	2	15	3 (18)
Managers	14	3	4 (7)	13	4	3 (7)
Systems designers	13	4	3 (7)	18	—	4 (4)

Figure 3. The strength and extent of values in each organization (shown by the number of X and Y modes)

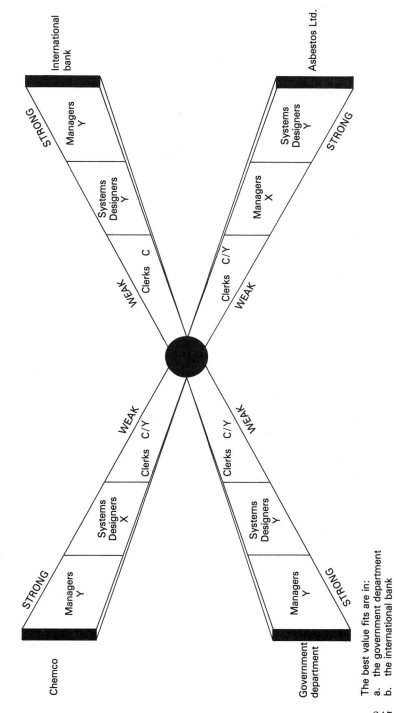

The best value fits are in:
a. the government department
b. the international bank

247

Table 2. The value fit

Government department

Good fit on values between all three groups	Computer system approved of. System designed to maintain and increase job satisfaction, and succeeded in doing so.

International bank

Good fit on values between clerks and managers and systems designer responsible for design of on-line system.	Computer system approved of. Initiative to associate new system with new form of work organization came from management and systems designer. Job satisfaction affected by later grading, staffing and promotion problems.

Asbestos Ltd.

Good fit on values between clerks and systems designers. Less good fit between these two groups and management.	Computer system approved of. Initiative to associate new system with new form of work organization came from systems designers. System improved job satisfaction.

Chemco

Good fit on values between clerks and management. Less good fit between these two groups and the systems designers.	Home sales computer a success. Export sales less successful.

because this relationship seemed to exist in this research it would necessarily be true in other organizations. Responses to change are a result of a combination of many variables, and although values are likely to be an important influence on how change is introduced and received, other factors may be equally important, for example power [Etzioni 1968: 319].

11.5.2. Work Values

The interviews with systems designers and managers included questions on what factors contributed to efficiency, happiness and a sense of fairness and justice in an organization, considering both role expectations (what was expected of staff) and role requirements (what the organization must provide). It was believed that the answers to these questions would throw some light on values through identifying the emphasis placed on cognitive factors, or the most efficient means to a given goal; cathectic factors which produce emotional responses of approval or disapproval, or moral factors associated with the evaluation of what is right or wrong, fair or unfair. The hypothesis was that those individuals or groups holding Y values would suggest more cathectic factors than those with X values who would focus on cognitive and instrumental factors. This hypothesis was not entirely correct as the principal difference lay between the government department and the industrial and commercial organizations, rather than between X and Y values.

Table 3. Cognitive, cathectic and moral values*
(in percentages)

Government department			Asbestos Ltd.			International bank			Chemco		
Cog.	Cath.	Moral	Cog.	Cath.	Moral	Cog.	Cath.	Moral	Cog.	Cath.	Moral
33	28	38	44	28	28	39	36	25	39	39	28

These figures are based on the number of *different* suggestions made. N = 216. If more than one person made the same suggestion it was only included once.
* Please see the case study chapters for the lists of suggestions.

The emphasis in the government department is on the contribution moral and cognitive factors make to efficiency, happiness and a sense of fairness and justice. In the other three organizations it is on cognitive and cathectic factors. The industrial and commercial organizations had cathectic expectations of their staff, expecting them to hold certain attitudes such as commitment, enthusiasm, motivation. This was less true of the government department.[3] The ranking of each organization is shown in Table 4.

Table 4. Ranking of value orientations

	Moral/cognitive orientation			Cognitive/cathectic orientation	
Government dept.	38%	33%	Government dept.	33%	28%
Asbestos Ltd.	28%	44%	Asbestos Ltd.	44%	28%
Chemco	28%	39%	Chemco	39%	39%
International bank	28%	39%	International bank	39%	36%

---

3. This was also the case with the other two government departments, where only 19% and 24% respectively of answers were in the category.

Too much weight should not be put on these figures as they are not statistically significant but they suggest that the concept of equity is very important in the government department both in relation to its own staff and to the public. It will be recalled that a major influence on the design of the government department computer system was the need to treat the public in an equitable manner. In contrast, the other organizations, the international bank and Chemco more than Asbestos Ltd., placed considerable emphasis on developing a set of pleasant and caring relationships with their staff.

Lastly, in order to identify the values of managers and systems designers in relation to the systems design process, each group was asked to list and rank what he would include in the phrase systems design. Each group answered in terms of his own area of interest and 63% of the points made by systems designers referred to systems design procedures, whereas 78% of the activities in the managers' lists were concerned with the identification of their business needs and the requirements for successful implementation. This would suggest that the most managerially successful systems will be those in which managers and systems designers complement each other by working closely together. However, management-systems designer collaboration was not a problem in this research. In the government department, systems designers had been managers and user management was represented in the design of each system. In the two firms and the bank there was close cooperation between the two groups.

Cohen et al. [1972: 4] have defined organizations as ''a collection of choices looking for problems, issues and feelings looking for decision situations in which they might be aired, solutions looking for issues to which they might be the answer, and decision makers looking for work''. It would be useful to add ''values looking for choices and actions through which they can make themselves explicit''. Etzioni has said: ''most societies are short not on values but on their realization'' [1968: 13] and that ''values which are not mediated through concrete social structures tend to become tenuous, frail and, in the long run, insupportable''.

What people say their values are is a useful guide to their behaviour but values will tend to become clearer and more specific as change is implemented and the actual consequences of a decision become visible [ibid: 265]. Also, just as values have influenced the choices made, so the consequences of these choices will in turn affect values.

The data presented in this section have shown that there were similarities in values and value relationships in our four organizations and in the next section we can see how these influenced the methods used for attaining goals.

The data have also identified differences in the perspective of systems designers and managers and these too may influence the goal attainment process.

250

11.5.3. Conclusions on Values and the Change Processes

Successful change requires a number of different processes to be successfully accomplished. Goals and objectives have to be set and attained, an existing complex socio-technical system has to be transformed into a different and perhaps more complex system, equilibrium has to be restored so that the new system can operate effectively and this state of equilibrium has to be maintained into the future, while the system accepts and responds to further change. Parsons has called these different aspects of change "goal attainment", "adaptation", "integration" and "pattern maintenance" [1964a: 65; 1964b: 330]. Although seeming to be located on a time continuum, in reality they are likely to occur simultaneously, overlap, or form different sequences with one subsystem being at one stage and a second at another.

All change processes involve the use of politics and power, the latter being defined as an ability to exert control over scarce resources so that outcomes can be produced which are consonant with an individual's or group's perceived interests [Mumford and Pettigrew 1975: 205; Salancik 1977]. These scarce resources may include expertise, information, political access, prestige, and the support of others [Pettigrew 1968, 1973a]. But because power derives from activities rather than individuals, an individual or subgroup's power is never absolute and derives ultimately from the context of the situation [Hickson 1971]. The study of politics and power, although crucial to an understanding of the processes of change, is only possible if the researcher is able to be in a research situation over a long period. This was not feasible in the present research and so this aspect of change will not be discussed.

11.5.3.1. *Goal Attainment*

Change provides an opportunity for making choices and the kinds of choices that are made provide an insight into the values of the different groups involved in the change process. For example, values may lead to problems being defined in such a way that choice opportunities are limited [Christensen 1974] or conflicts of values and interests may give the choice process a dynamic which makes outcomes difficult to predict. Again the fact that choices are made and goals are set does not necessarily mean that these goals are achieved. They may not be possible to achieve, the processes of implementation may distort them or new circumstances may make goals set at an early stage, irrelevant at a later stage. The goal attainment process will broadly follow the stages shown in Figure 4, although in a much more confused and less clear-cut manner.

Values will not be the only influence on the kinds of goals that are set, these will also be constrained by the nature of the decision-making environment. This will include the number and diversity of the groups involved, the efficiency of information systems in providing the decision group with relevant, broad-based, and up-to-date information and the extent to which internal and external organizational environments are subject to change [Friedman 1967].

Not all planned change incorporates the setting of goals; in some very uncer-

Figure 4. Stages in goal attainment

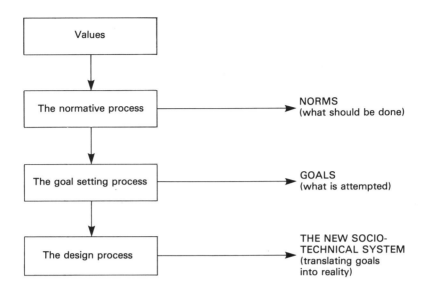

tain situations goals may prove a restriction rather than a stimulus [McCaskey 1974], but all the four organizations in this research claimed to have set very precise goals at the start of their projects. The majority of these goals were cognitive rather than cathectic or moral and directed at the attainment of economic, rational or technical objectives. The intellectual model used to attain goals was close to that recommended by Steiner.

> Planning is a process that begins with objectives; defines strategies, policies and detailed plans to achieve them; which establishes an organization to implement decisions and includes a review of performance and feedback to introduce a new planning cycle. [1969: 27]

This was particularly true of the government department whose extremely large system involved the setting up of a network of groups to take and implement decisions. The formulation of clear objectives appeared to give a sense of order, clarity and purpose to the design process although it may have had the disadvantage of inhibiting the exploration of alternative futures once the movement towards goals had begun [McCaskey 1974].

Parsons has described goal attainment as a directional change that tends to reduce the discrepancy between the needs of the system with respect to input-output interchange and the conditions in the environment that bear upon such

252

needs [1965: 39]. Listed in Table 5 are the principal goals set by the four organizations, the environmental influence that stimulated the goal and an assessment of whether the goal was successfully achieved.

Table 5. Set goals, environmental influences and attainment of goals

Goal	Environmental influence	Attainment of goal
Cognitive goals		
Economic Reduce staff numbers	*External* Tight labour market meant staff difficult to recruit and costly	Yes. Considerable staff savings achieved in government department. Some savings elsewhere
Economic System must pay for itself	*Internal* Treasury and Boards of Directors requirement	Yes, in government department. Probably yes elsewhere
Rational Increase efficiency	*External* Customer requirement for faster, more accurate service	Yes, orders, claims, etc., processed more quickly
Rational Increase control	*Internal* Organizational requirement for better control	Yes, through better information
Technical Keep up to date with technical developments	New technical equipment on market	Yes, through designing line system
Cathectic goals	*Internal*	
Socio-technical Increase job satisfaction (3 organizations)	Management/system designer values. High labour turnover.	Yes, in 2 organizations
Socio-technical Remove man-machine barrier (1 organization)	User dislike of previous batch system + systems designer values	Yes
Moral goals Provide equitable service to public (in government department)	Organizational values + public pressure	Yes

253

All of these goals, with the exception of the economic goal that the system must pay for itself, were a response to some perceived discrepancy between the existing situation and the situation as an influential group in the organization would like it to be [Vickers 1973: 171]. This discrepancy acted as a stimulus for change [Golembiewski 1970: 18]. Dewey has described organizations as being in a constant state of unrest with goals as a "reflection upon what one will do" [Dewey 1957: 69]. The conscious tendency towards change or action has been described as conation and Dewey sees this as incorporating a number of steps. First the nature of the problem must be identified; second, information resulting from previous experience with similar problems must be evaluated. Finally a judgement must be made which synthesizes these present and past observations into a plan for action. This process must create unique goals as past circumstances will never exactly replicate themselves [Dewey 1957: 234]. This is particularly true of computer systems where the rate of technological change means that early equipment is replaced with more sophisticated hardware.

An interesting feature of the goal setting process observed in this research was that technical goals did not dominate. Sackman has described very critically how the programmers and systems analysts of the 1950s and 1960s were technically blinkered. He says: "Most programmers tended to view accommodation for users as costly and disconcerting in the more urgent business of building programs and making them work with limited resources" [1971: 47].

Speaking of computer science curricula today he says:

> They are oriented toward mathematical and logical prowess rather than toward a sensitivity to real world problems and social responsibility. In the pursuit of excellence the computer world aspires towards technical heroes – there are no humanistic heroes. [ibid: 49]

In contrast to this technical perspective, the strongest influence in the four organizations was the wish to achieve what were seen as important business requirements; the reduction of costs and the improvement of efficiency. In those organizations which set social or socio-technical goals, the needs of user management and clerks were given high priority, although not at the expense of efficiency. This business and social orientation was a consequence of a number of factors. First, computer systems are easier to design today and therefore technical constraints are fewer. Sackman states: "Early man-computer communications were closely tied to the constraints of machine circuitry, and were tightly bound to logical and mathematical symbology" [ibid: 46].

Second, the power situation has altered as a result of a wider spread of knowledge through increased use of computers. Effectively computer specialists have lost their monopoly of EDP information [Mechanic 1962; Mumford and Pettigrew 1975: 199]. In the government department application the systems designers were recruited from staff who had previously worked as administrators in the departments which were to be computerized, and who therefore had considerable knowledge of and sympathy for the needs of the user, at least at managerial level. The two firms and the bank were introducing their second stage of computerization and the systems designers were all conscious that their design responsibility covered the setting and meeting of business as well as technical goals. They therefore worked in close association with user management throughout the design process.

Another feature of the goal setting process was that the goals were all attainable, and all attained, suggesting perhaps that innovators who are going to be later judged on the success of their actions do not make explicit objectives which are difficult, uncertain, and may not be achieved. An alternative explanation is that because the author was dealing in history, although recent history, successes were remembered and failures forgotten, thus preserving the psychological security of the reporters.

The principal features of the goal setting process can be summarized as follows:

(1)   Goals did reflect organizational values and were associated with the attainment of business and human rather than technical objectives.

(2)   A number of important goals represented the values of organizational subgroups. For example, the socio-technical goals related to increasing job satisfaction originated with management in the government department and the international bank, and with the systems designers in Asbestos Ltd.

(3)   Those who participated directly in the early stages of the systems design process, for example, user management, were able to influence the nature of the goals set.

(4)   The goals that were formally set and made explicit were those that could be attained without difficulty. This enabled the systems to be seen as a "success" when they were evaluated after implementation.

The principal vehicle for the attainment of goals was the systems design process. Many books have been written about this and so it will be dealt with only briefly here [Mumford 1964, 1968, 1969, 1971, 1972, 1975a, 1975b]. It has been argued that the choices individuals and groups make reflect their values; therefore the technical and human design alternatives that were considered by the four organizations are important sources of information on values. But it seemed that few technical design alternatives were examined. The early batch systems were designed to fit existing manual systems and the subsequent on-line systems were an enhancement of the batch systems. One reason for this limited search process was a desire to avoid unnecessary disturbance, either to the public in the government department or to staff in the two firms and the bank, where offices were working reasonably efficiently in a manual mode. Technical choice associated with the on-line systems was primarily,

(1)   whether to have an on-line/batch combination, or on-line/partial real time, or on-line/total real time system;

(2)   how to handle input and output;

(3)   how much information to provide via the VDUs.

Technical choice was therefore constrained by existing systems and by a desire to introduce change without causing too much disturbance. Both of which inhibit-

255

ed a radical thinking of how to solve problems. This approach fits Rosenbrock's description of the design process as a compromise involving many trade-offs [1975: 4] and Hedberg's suggestion that design choices have to be based on belief rather than knowledge of the instrumentality of actions for desired outcomes [1979].

The choices associated with the design of the social part of the systems are of more interest. When organizations do not set any positive human goals the reasons given for not doing so are usually of the following kind.

(1) Job satisfaction is not the responsibility of the systems design team. It is the responsibility of the user;
(2)   Human goals should be related to the needs of the public, not to the needs of staff (a government department viewpoint);
(3)   Managerial goals such as efficiency and security must be given priority. Human goals such as greater work flexibility could threaten the attainment of these;
(4)   Clerks are compliant and unlikely to protest (an EDP specialist's viewpoint);
(5)   Clerks are resources facilitating the operation of a technical system.

Hedberg has said:

> The actors in the design process model their perceived reality in mental and formal models, and they use these models to arrive at plausible design alternatives . . . . The opportunities for situational learning are few, since few designers repeatedly construct similar systems in relatively stable environments. [ibid: 11]

These systems designers appeared to hold a very simple mental model of the clerks for whom they were designing systems, viewing them as compliant resources [Hedberg 1976: 62]. This would seem to be a high risk strategy for as Friedmann has said:

> The views of those on whom the functioning of a system depends cannot be wholly ignored unless the system is prepared to part with their services – in which case it must come to terms with their replacements. [1967]

The approach in the three organizations which set specific human goals related to job satisfaction, the government department, the international bank and Asbestos Ltd., was very different. In each of these situations great thought was given to different ways of organizing work and to the creation of interesting and challenging jobs. In the government department this was done entirely by the systems designers and management as the clerical staff were dispersed around Scotland and had not yet been transferred to the new office centre. In Asbestos Ltd., design alternatives were worked out by the systems designers assisted by the author and with considerable consultation with the clerks. The knowledge acquired by the clerks during this process led to them rejecting the job enrichment solution recommended by the systems designers and substituting their own, based on the creation of autonomous groups. In the international bank total responsibility for the design of the social system was given to a representative group of user clerks who also chose an autonomous group form of organization.

The fact that social as well as technical goals had been set had two consequences for the systems designers. First, it made the systems design process more complex, requiring a knowledge of social as well as technical systems design. Second, it introduced the constraint that technical design had to be very flexible so that the technical part of the system did not inhibit any desired social solution. Three conclusions can be drawn from an examination of the consideration given to human design alternatives. These are that,

(1) if a human goal such as an increase in job satisfaction was accepted and made explicit then considerable efforts would be made to attain it;
(2) If no precise human goals were set then little attention would be paid to human needs in the systems design process;
(3) if the user, particularly the user clerk, participated in or controlled the design of the social system then goals which were important to users, such as job satisfaction, were likely to be attained.

11.5.3.2. *Adaptation*

The process of adaptation is concerned with moving from one state of equilibrium or integration to another and the means by which this is assisted to take place smoothly and easily. Rapid alteration without stress does not readily happen of its own accord and there is a need for philosophies, structures and strategies to assist the process. These will include values, coalitions, task groups, incentives and mechanisms for avoiding or resolving conflict for, like goal attainment, adaptation requires the reconciliation of different interests and objectives.

Adaptation will be easier for some groups than for others as they may feel less threatened by the change, they may be more change-minded and accustomed to coping with new situations, or they may have power and so be able to mould the change situation to their own interests. Adaptation is therefore a political and negotiating process in which a compromise rather than an optimal solution is likely to be achieved, although the aim should be the easy and rapid adaptation of all the groups involved. The author was not able to be present when the systems in two of the four organizations were being designed and implemented and so she is not able to identify the influence of political factors. This analysis will therefore be restricted to adaptation structures and strategies as observed by her or described by the systems designers and user managers.

11.5.3.2.1. *Adaptation Structures*

Any major change process is a test of an organization's philosophy. It will show whether the philosophy had adequately prepared the organization for change, and if it assists the successful management of change [Mann and Williams 1960]. The organization has to be capable of introducing and legitimizing new objectives, procedures and attitudes and of creating a structure and a set of levers to enable it to do this [Friedmann 1967]. It requires what Gross has called an ''institutionalized capacity to build other institutions" which in turn requires "the development and maintenance of a network of supporting groups" [Gross 1967: 201]. Edstrom has suggested that when large-scale computer systems are being designed there will be a great deal of top management intervention for this group will wish to influence the outcome of a change that is going to have considerable organizational impact [1974].

The structures set up to assist adaptation in our four organizations are shown in Table 6. It can be seen that Edstrom's hypothesis is supported, for the government department had a senior management steering committee whose role was the formulation of policy and the evaluation of proposals from the systems design groups. It also set up a liaison committee with the objective of bringing together at senior level all the different interest groups, including the Staff Association and the Inland Revenue Board. This approach is in line with Gross's belief that successful large-scale change requires coalitions which include "key activators", "active allies" and "passive collaborators". Passive collaborators are those groups who do little more than "go along" with change but whose acquiescence or neutrality is of tremendous importance [Gross 1967: 202]. Asbestos Ltd. also had a steering committee but this had the very different role of advising on the participative structures and strategies in the user department. It was not concerned with top level policy formulation.

Each organization was careful to involve its user management in all aspects of the change process. In the government department, management was an influential part of the design team and so adaptation came as a result of direct participation in system development. In Asbestos Ltd., Chemco and the bank every effort was made to design the kind of system that management wanted and this policy stimulated managerial enthusiasm for the change. Managerial participation was formalized through the setting up of user committees with responsibility for developing new administrative procedures to fit the new level of technology. These efforts to involve management were in sharp contrast with the situation ten years ago when the author was studying the introduction of computer systems in other organizations. A common policy then was for the EDP group to design a system and "sell" it to management. It was believed that management's ignorance of computers meant that it had nothing to contribute to systems design [Mumford and Pettigrew 1975].

But, whereas participation at management level was considerable, this was only true of clerks in those organizations which had set job satisfaction goals. Chemco tried to consult with its clerks but found this difficult because of their lack of involvement and knowledge. The government department used its Whitley structure extensively as a sounding board once the system was implemented but had the problem that prior to this there were no staff to consult with. In contrast, Asbestos Ltd. and the international bank created committee structures which enabled their clerks to assume design and implementation roles.

The creation of a support network of committees did appear to assist adaptation in the four organizations. These committees provided opportunities for influence, for negotiation, for learning, and for taking responsibility. They worked best

258

Table 6. Adaptation structures

Adaptation structure	Organizational level of membership	Adaptation function	
Government department			
Steering committee	Senior management	Change policy	
Liaison committee	Senior systems designers, users, TU, Board	Change policy	
User committees	Management	Development of new procedures	
Design groups with user members	ADP, management	System transformation	
Consultative (Whitley) committee	Clerks	Problem identification once system was implemented	
Industry and bank			*Used by:*
Steering committee	Systems designers, user management, TU, personnel department	Change policy and progress	Asbestos Ltd.
Progress committee	EDP, management	Decide priorities, monitor progress	Chemco, international bank
User committees	EDP, management, clerks	Development of new procedures	Asbestos Ltd., Chemco
Job satisfaction working parties	Clerks	Design of work organization	Asbestos Ltd, international bank
Implementation working parties	Clerks	System transformation	Asbestos Ltd., international bank

when the members were not restricted to a passive role of question or comment but were given specific tasks so that they could actively contribute to policy or design. Each organization formed groups of this kind at management level and seemed to appreciate the point made by Friedman about the relationship of experts and politicians in national planning.

> An efficient decision process almost always involves both experts and politicians [managers] simultaneously in close interdependency. No politician [manager] who values the services of an expert can afford consistently to disregard his judgement, nor will any expert desiring influence, systematically oppose the wishes of his employer. [1967]

When clerks are excluded from committees this may be due to management's view of them as a compliant group with little interest in what is taking place. But, as Gross has pointed out, "the support of subordinates cannot be taken for granted. It must be won and maintained through specific actions taken by their leaders" [1967: 204]. Alternatively, there may have been a belief that successful coalitions of technical designers and user managers would be threatened if a third group was introduced. If so there was a lack of awareness of the concept of "conditionality" described by Weick [1967: 37]. This states that the relationship between two components is conditioned by the state of a third component. If these components are systems designers, managers and clerks then it is the relationship between all three groups that will control the system.

11.5.3.2.2. *Change Strategies*

There are many strategies which one group can use to ensure the compliance or cooperation of another. There is compliance through deference, because a group knows its place; or through trust because there is confidence that the right thing will be done. There is compliance through the understanding which comes from good communication. There is compliance through negotiation and through shared control and participation in decisions. Which of these strategies is used will be determined by power, by expediency and by values. Table 7 shows the strategies used in the four organizations.

Table 7. Adaptation strategies

	Communication with	Consultation with	Participation in decisions	Control by
Government department	All groups	User m/ment, staff reps.	User m/ment, staff reps.	User management
Chemco	All groups	User m/ment, section leaders	User management	ADP/user management
Asbestos Ltd.	All groups	All groups	All groups	ADP/user management
International bank	All groups	All groups	All groups (clerks through representatives)	All groups (of some aspect of design)

All the organizations used the strategy described by Sayles as "show and explain what is intended so that employees can understand and thereby become reassured" [1974: 191]. Sayles suggests that this approach, if used on its own, assumes an omniscient giver or donor culture impacting a backward impotent donee culture. If the giver is benevolent he endeavours to soften the blow and aid the donee to make the best use of this new technology or procedure. A similar approach is what Schon has called the "centre-periphery model" which has the following components:

(*a*) the innovation to be diffused exists, fully realized in its essentials, prior to its diffusion,

(*b*) diffusion is the (centrally managed) movement of an innovation from a centre out to its ultimate users, by dissemination, training and provision of resources and incentives [1971: 81].

But good communication is an essential part of the most successful change strategies and our organization also consulted with user managers and involved them in decision taking. This was a wise policy since there is evidence that user and technical knowledge must be blended together throughout the development cycle if change is to be successful. Sayles describes how NASA's early technology utilization programme had only limited success because it wrongly assumed that if a potential user was informed that a new technology existed, then he would use it [1974: 198].

There are four arguments for involving lower level groups such as clerks in a more fundamental way than merely communicating to them what is going to happen. The first is a values argument which states that people have a moral right to control their own destinies and that this applies as much in the work situation as elsewhere [Mumford 1978]. The second is an expediency argument and states that activities are ultimately controlled by those who perform them, and that people who do not have a say in decisions may decide to repeal the decisions of others as soon as those others leave the scene [Hedberg 1976; Nystrom 1975]. The third relates to the location of knowledge and states that the experts on operational factors such as task design are the people who do the jobs [Edstrom 1974]. The fourth argument is that involvement acts as a motivator [White and Ruh 1973]. Asbestos Ltd. and the international bank were probably most influenced by the third argument, with the first also carrying some weight, the government department by the first and fourth arguments. Chemco, which involved user managers but not clerks, by the third argument. The international bank handed over more control of design activities to the clerks than Asbestos Ltd., although the Asbestos clerks were able to influence the choice of organizational structure for their department.

This study of adaptation processes in the four organizations suggests that the development of structure and strategies to assist adaptation is worthy of more thought than is often given to this aspect of change. Sayles suggests that most of the really difficult problems of innovation occur in the middle, not at the beginning or at the end. Yet most academic attention has been given to these "end points" through studies of creativity or the acceptance of innovation [Sayles 1974]. Successful adaptation requires a careful diagnosis of the needs of the groups involved in the change. This diagnosis should cover both communication and consultation strategies for each group, and the best distribution of responsibility and control. In order to implement selected strategies, the right kind of structure needs to be created. The values of the author lead her to suggest that a

261

successful method for identifying appropriate structures and strategies is for each group to assume responsibility for identifying its own needs, the role it wishes to play in the change process and the degree of responsibility and control it wishes to assume. Inevitably this will lead to conflicts of interest and so a higher level system, such as a steering committee, may be a useful mediator.

An important function of adaptation structures and strategies is to reduce the ambiguity and equivocality in the change process and to give people a sense of control, knowledge and purpose; an understanding of where they are going and what they are trying to achieve. A number of writers have suggested that an absence of equivocality is an important factor in an individual or group's satisfaction [Weick 1969: 99; White 1959]. Byrne and Clore have called this a desire for "effectance".

> Any situation which provides evidence of one's predictive accuracy, ability to understand, correctness, logicality, reality orientation, behavioural appropriateness – any information which permits or indicates effective functioning – would satisfy the effectance motive. [1967: 4]

The four organizations created a comprehensive network of groups to assist adaptation, whose roles varied from the creation of policy to responsibility for specific tasks. An important question is how rigid or fluid such a group structure should be. Again the answer is likely to depend on the needs of the situation and on whether these needs are political or problem solving. Groups which wish to protect themselves against certain aspects of change, may prefer a stable and permanent structure that enables them to constantly monitor what is happening and ensure that this is in line with their interests. But a complex change will also require more fluid groups in which knowledgeable individuals come together to tackle particular problems and disband once the problem is solved. An organizational model which enjoys some academic popularity at present is called the self-designing organization. Hedberg and others have described it in the following terms:

> Expertise should be diluted, authority ambiguous, statuses inconsistent, responsibilities overlapping, activities mutually competitive, rules volatile, decision criteria varying, communication networks amorphous . . . . [1967]

This model does not remove ambiguity and equivocality; it deliberately builds it in.

The best way of assisting adaptation probably lies in good diagnosis and experiment. A dynamic situation requires rapid feedback on which structures and strategies are working well and badly as events stream through the change processes. Ultimately those selected will reflect organizational values, and change provides an excellent opportunity for making these explicit.

262

11.5.3.4. *Integration*

Adaptation has been defined as the movement from one integrated set of relationships to another. Integration is when a number of parts of a system are mutually adapted to each other and contributing to the successful functioning of the system as a whole. These parts or units can be people or things or a combination of the two as in the concept of a socio-technical system [Bredemeier 1962: 54; Neal 1965: 12]. Parsons sees integration and how this is achieved as central to the interests of the sociologists [1961: 40]. His theories are strongly oriented towards stability, value consensus and harmony of interests. He perceives change as the process of moving essentially stable systems from one state of integration to another. Change occurs because a system gets out of phase with its environment and needs to move to a new state of balance. These views of Parsons have been strongly criticized. C. Wright Mills has referred to them as "deviant assumptions" that "impede the clear formulation of social problems" [Mills 1959: 41]. The Marxists view different groups in society as fundamentally in conflict and coherence and order a consequence of implicit force and subjugation.

Integration is made difficult by the fact that organizations are complex entities that function on several levels [Touraine 1971: 147]. Touraine tells us that employees act simultaneously in terms of their own interests, their status, their roles in the organization and the power conflicts in which they are involved, while the firm has to integrate the different levels of operational tasks, organizational structure and management objectives and mediate between different strategic and political interests at each level. Some writers have argued that integration was possible when employees did not question the logic of bureaucratic organizations in which they worked. But today, not only are organizations rational, rational habits of mind have also infected the work-force and they are now questioning what they previously accepted [Fitzgerald 1971]. Others believe that integration has never been possible. That there are no social systems that can be perfectly integrated with their cultural systems and with the personality systems of their members; that the values institutionalized in a social system will always be somewhat inappropriate to some parts of that social system [Menzies 1976: 117]. Gowler suggests that too tight an integration is not even desirable as it removes the stimulus for change [1972].

Some organizations appear to be able to survive with a high level of conflict and a low level of integration, but in most the managerial task is seen as bringing together the different parts into a well-functioning whole that is both technically efficient and satisfying in human terms. At any one moment in time some parts are likely to be both inefficient and unsatisfying, nevertheless the managerial objective is what can be called "dynamic stability". A set of relationships which work well together yet which can be modified when there is a need for change. One way of securing these effective relationships is through reciprocity, the acquisition or meeting of mutual obligations [Blau and Scott 1963: 134]. The employment contract is one example of this kind of exchange relationship [Mumford 1972; Morley 1974].

When technical change has been introduced, successful integration requires the bringing together of people, technology and structure into a viable and stable relationship [Leavitt 1958: 323]. This has to be done within the context of an organization with values and goals as shown in Figure 5.

Figure 5. Successful integration of technical change

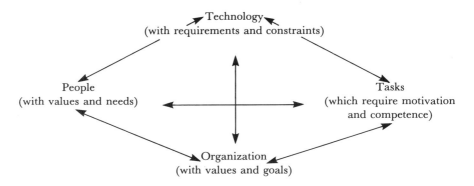

The relationship between these four variables needs to be stable but it should not be static. Organizations should be able to respond to new pressures from their environments while at the same time maintaining a state of equilibrium or being able to make adjustments which restore equilibrium if internal relationships are badly disturbed. The introduction of a new computer system is likely to affect each of the variables in the Figure. A new level of technology will bring with it a new man-machine relationship incorporating both opportunities and constraints. Because tasks are altered by technology, the task structure of departments using the computer will be affected. New tasks mean that new demands will be made of people and this will affect job satisfaction positively or negatively depending upon whether the new situation meets their values of what is desirable work. In turn, the technology, tasks and people variables will interact with an internal organizational environment which provides a structural context for the achievement of the organization's purpose and this interaction may start the looping process again by making new demands of technology. A lack of recognition of the importance of integrating these variables when introducing computer systems is seen as an important source of their failure [Bostrom and Heinen 1977]. Lucas, summarizing the results of his research into the success of information systems in sixteen organizations says: "It is our contention that the major reason most information systems have failed is that we have ignored organizational behaviour problems in the design and operation of computer-based information systems" [1975: 6].

In this research the author has concentrated on examining two aspects of integration. The first is the extent to which technology and tasks have been brought together into a set of interesting and stimulating activities which meet the criteria

264

associated with good job design. The second is how well this combination of technology and tasks is integrated with the values and needs of employees and produces feelings of job satisfaction.

11.5.3.4.1. *Technology, Values and Task Structure*

There is now considerable evidence that most employees prefer jobs that provide an opportunity for the use of discretion and that the exercise of discretion gives them a feeling of personal identity and responsibility [Fox 1974: 30; Wedderburn and Crompton 1972: 226; Shepard 1965: 117]. In contrast routine jobs which do not provide opportunities for the use of discretion are believed to induce a sense of worthlessness or anonymity [Fox 1974: 64; Selznick 1969: 188; Gouldner 1971: 277]. In this study only 22% of all clerks thought that their employers should "try to organize work activities into tightly structured jobs which are clearly defined and do not permit a great deal of individual discretion". The majority (39%) were on the Y side of the scale, preferring work flexibility, and the remainder in the centre. Two of the four organizations, Asbestos Ltd. and the government department, had set up autonomous group structures which allowed their staff the use of considerable discretion, together with the opportunity for looking after the needs of a specific group of customers or clients. A third, the international bank, had given its staff the opportunity to create the kind of structure which best fitted their needs, and the staff chose autonomous groups. If the hypothesized relationship between values and action is correct then the values of the managers and systems designers in these three organizations ought to support this kind of work structure. This can be tested by examining their answers to the questions on work structure on the value scales and these are shown in Figure 6. Chemco has been excluded from the analysis, because, although its management subscribed intellectually to flexible jobs which allowed the use of discretion, the firm had not introduced these at the lower levels of the clerical hierarchy at the time of this research. In order to provide comparative data the values of managers and systems designers in the two government departments which were studied but whose case studies are not included in this book, are shown. These government departments did not make job satisfaction a system objective or use a participative approach.

It can be seen that the hypothesis is supported. No managers and systems designers in the two other government departments are on the Y part of the scale on the left and centre charts, and only 9% on the right hand chart. In contrast a majority of the systems designers and managers in Asbestos Ltd., Inland Revenue and the international bank are on the Y side on two of the three questions. Fox [ibid: 42] suggests that the choice of task structure made depends upon the values of the manager, his priorities and the costs and benefits as he sees them. He says:

> A manager who is himself called punitively to account may limit the discretion of subordinates to a degree which encroaches on the adaptiveness and creativity of his department, preferring this to the threat of more visible disasters which may stem from too much autonomy.

265

Figure 6. The values of the managers and systems designers in Asbestos Ltd., the government department (Inland Revenue) and the international bank compared with those in two other government departments which did not make job satisfaction a system objective, or use a participative approach.

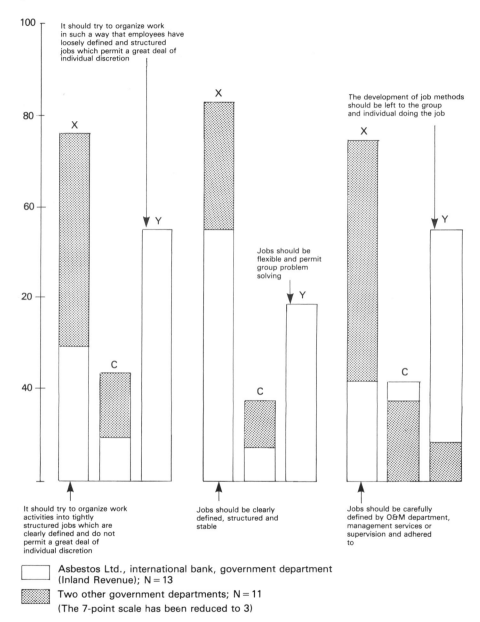

Asbestos Ltd., international bank, government department (Inland Revenue); N = 13

Two other government departments; N = 11

(The 7-point scale has been reduced to 3)

The managers of the two government departments which favoured a theory X approach argued that standardized work procedures and tight controls were essential because of their legal requirement to provide a service to the public and because of their anxiety to provide a service that was efficient and equitable. The Inland Revenue management, however, believed that it could still provide a good level of service and maintain efficiency using a more flexible team-based form of organizational structure. This allowed staff the use of judgement and discretion in their interpretation of the tax laws and of how to assist the public with its tax problems.

Fox has pointed out that what is being discussed here is the degree of confidence and trust between management and subordinates. Routine jobs with low discretion imply that the manager believes that his subordinates cannot be trusted, of their own volition, to perform according to the organization's goals and values. He also points out that trust is a reciprocal relationship. A management that does not trust its employees will not be trusted by them [ibid: 66]. Unless a socio-technical approach is taken to the design of computer systems and careful thought given to developing the kind of relationship between technology and work organization that provides staff with discretion and control, then computer systems can have a number of undesirable human results. They can increase the routinization of work by adding new routine tasks such as coding, checking and punching; they can remove activities which staff may have enjoyed such as interest, sales order and invoice calculations, and they can split a previously integrated job into the two halves of input and output [Mumford and Banks 1967; Whisler 1970]. The job structure comparisons in Table 8 show the considerable variety of task activities available with the form of work organization adopted in the government department and Asbestos Ltd. Although on-line computer systems fit well into this kind of autonomous group structure, for each group can have its own VDU, the government department's experience shows that autonomous groups can be associated with batch systems. Studies funded by the United States National Science Foundation on satisfaction and productivity suggest that the restructuring of work can improve both of these [Strivastra 1975; Katzell and Yankelovich 1975]. The important link in this improvement is task variety, information feedback, work related communication among employees, participation in decision making and technical characteristics of the job [Taylor 1976]. The Inland Revenue and Asbestos Ltd. group structures catered for all of these.

Davis sees us as entering what he calls the post-industrial era in which new needs and values will lead to new organizational structures and designs for jobs. He sees a change in cultural values from achievement, self-control, independence and endurance of distress to self-actualization, self-expression, interdependence and a capacity for joy. He says:

> The values of a society and its institutions slowly evolve in response to social and psychological conditions and to the conceptions the society holds of its environment. It is the emergence of some new values and the change in relative importance of others that leads to the recognition that we are witnessing the evolution of a new epoch, the post-industrial era. [Davis 1971].

Table 8. Job structure comparisons

			Variety				Knowledge choice			Job identity				Job relationships		Goal clarity	Anti-oscillation	Optimisation	Development	Overall control
			Reduce boredom	Give sense of personal control	Give feeling of achievement		Give sense of making important contribution			Give sense of team work	Give sense of confidence			Problem prevention	Efficiency improvement		Creativity	Autonomy	Key task

Control levels / Operational level headings (rotated):
No. of tasks · No. of skills · of methods · of work sequence · Use of judgement · Use of initiative · Clearly defined start to job · Clearly defined end to job · Uninterrupted task sequence · Long task cycle 20+ minutes · Visible contribution to product or service · Works as member of group · Considerable inter-action required · Clear work objectives · Objectives not too easy or too difficult · Can requisition required resources · Can correct errors, solve problems · Coordinates own work activities · Coordinates group work activities · Can improve methods · Can improve product or service · Individual free from supervisory control · Group free from supervisory control

Definition of skills:
A. Communicating in writing
B. Communicating verbally
C. Arithmetical skills
D. Machine operating skills
E. Checking/monitoring/correcting
F. Problem solving
G. Coordinating
H. Supervising

These are the day to day operating tasks

Inland revenue EDP batch

Role	No. of tasks	No. of skills	Notes	Key task
Tax officer higher grade – supervisor	L	6 A/B/E F/G/H	Long task cycle: with some cases	Problem solving, Organizing, Supervising
Tax officer	L	5 A/B/E F/G	Uninterrupted task sequence: for part of job; Long task cycle: with some cases	Problem solving, Data input and output

Asbestos Ltd. EDP on-line

Role	No. of tasks	No. of skills	Notes	Key task
Section supervisor	L	7 A/B D/E F/G/H	Optimisation: minor	Supervision, Problem solving
Territory clerk	L	6 A/B/D E/F/G		Order handling, problem solving

268

10.5.3.4.2. *Technology, Values and Job Satisfaction*

The kind of job an individual holds, and the tasks and responsibilities associated with this, influence his attitudes to work [Stone and Porter 1975], although these attitudes are mediated by the orientations and expectations which he brings to his job [Mumford 1972; Barrett 1970; Blauner 1960; Turner and Lawrence 1965; Vroom 1960, 1964]. This means that although employees may express dislike for certain aspects of their jobs such as routine, they may not report general dissatisfaction [Baker and Hansen 1975]. Three objectives of this research were, first, to ascertain the attitudes of the clerks in the different organizations to particular aspects of their work and to their work in general. Second, to establish if different ways of structuring work affected these attitudes. This would be done by comparing manual and computerized departments within the same organization, or the same department using first a batch and later an on-line computer system. Third, if possible, to establish the extent to which the introduction of different levels of computer technology affected attitudes and work.

The author has defined job satisfaction as the ''fit'' between job needs and positive job expectations (what the individual or group seeks and hopes to get from the job) and job requirements (what the job requires of the individual or group) [Mumford 1972]. She argues that a good fit between positive job expectations and job requirements leads to feelings of job satisfaction and she uses a modification of Parsons' pattern variables to provide her with a framework for measuring this fit (see chap. 3). The word ''positive'' is important here, for Gowler and Legge have argued that routine work can be acceptable to people with low expectations who have no experience of anything else. They say:

> Routinized and repetitive job performance, lack of autonomy and tight systems of organizational control all tend to deprive incumbents of the new experiences which challenge and arrest the corrosive effects of the old and familiar experiences which begin to dominate our thoughts, feelings and behaviours. [Gowler and Legge 1975]

This kind of negative role integration tends to come with age and long service when the option of moving to another job no longer exists; and because these employees do their work without complaint management may interpret apathy as satisfaction. It was not a characteristic of our four organizations where the majority of clerks were young. Positive job expectations should perhaps approach Argyris' view that the psychologically healthy person ''strives to be self-responsible, self-directed, self-motivated, aspires towards excellence in problem solving, strives to decrease his defensive and compulsive behaviour, to be fully functioning and so on'' [1964: 162]. Few work situations offer a majority of their employees these opportunities for self-actualization. For most people positive job expectations will be the attainment of interesting work in a pleasant and stimulating environment, with good pay and promotion prospects.

Locke has argued that satisfaction, dissatisfaction and other emotional reactions are value responses [Locke 1969, 1970]. We appraise an object or a situation against our standards of what we regard as good and beneficial. Gowler suggests that job satisfaction depends in part upon the individual's ability to achieve a degree of harmony between the affective, cognitive and evaluative elements in his personality (cathectic, cognitive and moral value orientations) [1974]. The work situation must give him some emotional satisfaction, stimulate his intellect and provide him with a sense of equity and justice.

Job satisfaction lacks theory but not measurement and many different techniques have been used to measure it. One method is to divide jobs up into a number of facets such as "variety", "opportunity to make choices", "control over pace of work", etc., and ask respondents to indicate how much of each facet their job contains; a subsequent question might be how important each facet item is to them. A second approach is to use what have been called discrepancy scores in which the respondent is asked not only how much of a facet there is in his job but also how much there should be, or how much he would like there to be [Seashore 1973, 1975]. A third approach is to obtain a measurement of overall satisfaction [Evans 1969]. This is usually done through attitudes scales, for example,

"Taking your job as a whole would you say that you enjoy your work, or not?"

. . . . . Enjoy it very much
. . . . . Quite enjoy it
. . . . . Do not enjoy it

The job satisfaction questionnaire used in this research incorporated all three measures (see Appendix B).

Although extensively used, measures of this kind still raise some doubts as to their validity. Taylor reports that irrespective of what measurement sophistication is brought to bear, 80% of Americans surveyed report being satisfied with their jobs [1976: 3]. Seashore argues that all job satisfaction surveys assume that there are some objective characteristics of a job to which people are responding. He believes that this may be true of some job aspects but not of all and that job satisfaction is a dynamic rather than a static concept in which some needs are subject to change. He also suggests that some responses may lie in the personality of the respondent rather than in his job; they are not based on objective reality [1973].

These criticisms are important and valid and the author's usual research approach is to feed back the data to the group who completed the questionnaires so that they can help uncover the meaning behind the statistics. Unfortunately, the large number of respondents involved in this project and the fact that the author carried out the research on her own, made this safeguard impossible. The following results should therefore be treated with caution.

As people may dislike particular facets of their work and yet be satisfied with the totality of their jobs, the best test of role integration, or the fit between what people are seeking from work and what they are receiving, would appear to be general measures of satisfaction. There were four of these in the questionnaire, formulated in the following way:

270

(A)   How often to you get a sense of achievement from your work?
(5 point scale: almost every day . . . . . hardly ever)
(B)   Taking your job as a whole, would you say that you enjoy your work, or not?
(3 point scale: enjoy it very much . . . do not enjoy it)
(C)   Some people are completely involved in their job – they are absorbed in it night and day. How involved do you feel in your job?
(5 point scale: very little involved . . . . . very strongly involved)
(D)   Think of a job you would regard as ideal from the work point of view. How does your present job compare with this?
(7 point scale: 7 representing ideal job).

Figure 7 shows the percentage of clerks who ticked the extreme points of the scales. The chart compares the percentages of clerks who can be viewed as having high job satisfaction in the manual and computerized sections of the Inland Revenue. It also compares batch with on-line systems in the Chemco export department and the international bank. The level of job satisfaction is also shown for Chemco home sales and Asbestos Ltd. The Chemco home sales and Asbestos Ltd. results are for their on-line systems.

It can be seen that in the government department, where the computer system had been designed to increase job satisfaction, it seemed to have had this result. In Chemco export sales there was little difference between the results of the survey carried out when the batch system was in use and the results with the on-line system, but job satisfaction was very high in the home sales section which used an on-line system. It will be recalled that in export sales both batch and on-line systems caused some frustration. In Asbestos Ltd. the fact that differently organized questionnaires were used to obtain opinion on the batch and on-line systems means that only attitudes to the on-line system can be shown on the chart. The author's view is that the on-line system, together with the new organization of work, contributed to an increase in job satisfaction. In the international bank both the on-line system and the reorganization into currency groups were approved of, but a number of personnel problems lowered morale. Staff expected that more varied and responsible work would lead to higher grades and initially some clerks were not given these. An attempt to secure staff savings by not replacing clerks who left caused overwork and stress, while the new system did not assist the long-standing problem of poor promotion opportunities. In addition a tight labour market for school-leavers meant that the bank could recruit extremely intelligent and well-qualified young people, yet work, although more varied than previously, was still essentially routine. Any fundamental job enrichment would require a loosening of bank controls and a form of organization that crossed departmental boundaries.

An interesting aspect of the results is the considerable differences in job satisfaction between the organizations, but this needs to be the subject of another book.

The percentage of clerks who were dissatisfied with their jobs is shown in Figure 8 and this can be regarded as the alienated group in each organization.

Figure 7. Percentages of clerks with high job satisfaction in manual compared with computerized offices and on-line compared with batch

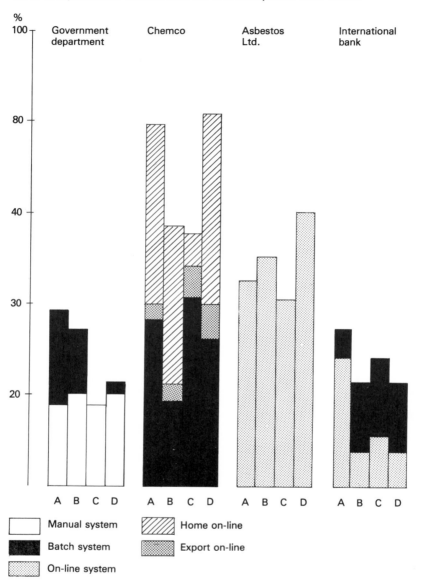

A = I get a sense of achievement almost every day
B = Taking my job as a whole I enjoy my work very much
C = I am strongly involved in my job (No clerks ticked the extreme category – I am very strongly involved; my work is the most absorbing interest in my life.)
D = This job is close to my ideal job (points 5 and 6 on the scale)

Figure 8. Percentages of clerks with low job satisfaction in manual compared with computerized offices and on-line compared with batch

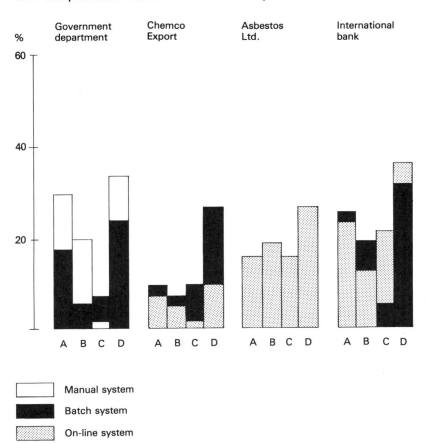

Manual system

Batch system

On-line system

A = I hardly ever get a sense of achievement
B = Taking my job as a whole I do not enjoy my work
C = I am very little involved in my job, my other interests are more absorbing
D = This job is far from my ideal job (points 1 and 2 on the scale)

In the government department there was more job dissatisfaction in the manual offices although very few clerks there did not feel involved with their work. In Chemco there was less job dissatisfaction with the on-line than with the batch work situation in export sales, and no job dissatisfaction at all in home sales. Asbestos Ltd., despite the new organization of work associated with the on-line system, still had some clerks who were not very happy; and while in the international bank fewer clerks using the on-line system said that they did not enjoy their work, this was offset by an increased percentage saying they felt very little involved in their job. These results appear to support conclusions reached by the author in previous research [Mumford 1969, 1971, 1973, 1974, 1975, 1978] that computers of themselves do not have good or bad human effects. The important thing is how they are combined with sets of tasks and responsibilities into an effective socio-technical system, and the extent to which this socio-technical system, and the environment surrounding it, meets the needs of the employees who operate it. In the government department, Asbestos Ltd. and the international bank, where the on-line computer systems and associated work organization were designed to increase job satisfaction, good socio-technical and role integration was achieved in the first two. The system needs to settle down in the bank before it can be evaluated. The Chemco home sales department had a very high level of job satisfaction although the extent to which this was assisted by the on-line computer system was not easy to ascertain. In contrast, in export sales a computer system which had not been easy to integrate into the work of the section in either its batch or on-line form caused some frustration.

Many writers have argued that job satisfaction is related to the amount of discretion in an individual's job and that successful role integration is assisted by the employee having the amount of discretion that he wants [Blauner 1964: 98; Fox 1974: chap. 1; Touraine 1955: 429]. Figure 9 shows the amount of discretion clerks in the four organizations said that they had, and also the amount that they said they would like to have.

The clerks in the government department and Asbestos Ltd. had most discretion and these were the two organizations which had autonomous group task structures. In view of the flexibility of the government department work environment and the skilled nature of the work it may seem surprising that overall job satisfaction measures were not higher. The explanation probably lies in the centralization policy being used at the time of the research. The large office complex and its remoteness from the public offset the advantages of a well-designed computer system and a flexible organization of work.

Successful integration therefore requires the bringing together of many variables into a state of harmony. Technology and work organization must reinforce each other in providing both efficiency and the kind of environment which stimulates feelings of job satisfaction. At the same time this socio-technical combination must meet the needs of the employees who form part of the system, and the whole must integrate well with external systems. The achievement of this integration requires careful diagnosis of technical, organizational, social and environmental needs and the creation of work systems that meet these complex sets of needs. But

Figure 9. Perceived discretion in work compared with desired discretion

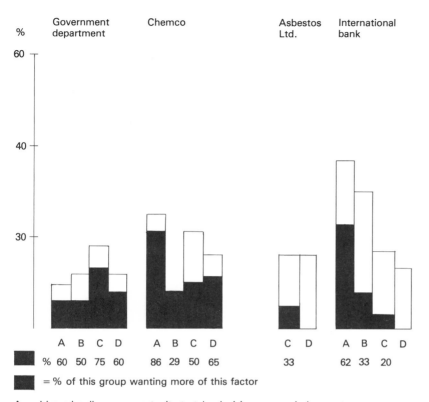

		A	B	C	D		A	B	C	D			C	D		A	B	C	D
■	%	60	50	75	60		86	29	50	65			33			62	33	20	

■ = % of this group wanting more of this factor

A = I have hardly any opportunity to take decisions or use judgement
   ■ I would like more opportunity to do so
B = I can hardly ever decide how to go about doing a job
   ■ I should like more freedom to do this
C = I hardly ever have a say in what I shall be doing next
   ■ I should like more choice
D = I almost always have to do things in a set order
   ■ I should like more freedom to choose the order of my work

this diagnosis and development will not take place unless people of influence believe that it is the right thing to do. This leads us back once again to the importance of humanistic values and to the words of Vickers:

> But as the rate of change quickens, whilst the rate at which the generations change becomes if anything slower, the need to change the standards we live by even while we use them, becomes even more important and even more threatening. The function of resetting norms and values becomes a conscious one. And with it we can discern a new level of control, a level of collective self-control or self-determination which casts special responsibility on the policy maker. [1973: 181]

11.5.3.5. *Pattern Maintenance*

Every social system is governed by a value system which specifies the nature of the system, its goals, and the means for attaining these goals. Smelser suggests that a "social system's first functional requirement is to preserve the integrity of the value system itself and to ensure that individuals conform to it. This involves socializing and educating individuals, as well as providing tension controlling mechanisms for handling and controlling individual disturbances relating to the values". This statement, placed in conjunction with that of Vickers above, presents us with today's dilemma and challenge [1959: 11]. The aspects of change that have been discussed in this book: goal attainment, adaptation, integration and pattern maintenance all require that the members of a society be more or less committed to them and agree on how they shall be achieved [Neal 1965: 11]. Pattern maintenance or the maintaining of a system in a state of integration and balance requires, according to Parsons, the "maintenance at the cultural level of the stability of institutionalized values". Yet the problem for the future is surely that of being able to hold societies together without violent insurrection or civil war, while at the same time assimulating and controlling the change in values desired by Vickers.

What will happen to our four organizations in the future remains to be seen. The British Civil Service, stimulated by the ideas of many farsighted individuals, and assisted by the importance which both management and unions have always placed on good human relations, appears to be placing great emphasis on humanistic values. The Inland Revenue was pioneering this new philosophy three years ago, but since that time many other humanistic experiments have taken place [Robertson 1973; Pace 1975; Burden 1976; Hunter and Crampton 1977]. In industry and commerce, Asbestos Ltd. and the international bank were leading the way in participative systems design when this research was undertaken, but many others are starting to follow their example. Chemco, a firm whose values have made it place great importance on policies to improve the well-being of its employees, is carrying out many quality-of-working-life experiments at the present time. Change of a humanistic kind is therefore accelerating. The question is: will it be fast enough and fundamental enough to meet the challenge

276

of an increasingly turbulent technical, economic and political environment in which many groups, with many conflicting demands, are seeking to be heard.

10.6. CONCLUSIONS

Parsons, whose ideas have provided the analytical tools for this book, was the prophet and proselytizer of stability. For him all societies and systems were inherently stable, although this stability had to be worked for. Today new prophets declare that the "stable state" was in fact a myth and that the future is one of increasingly fast and increasingly complex change in technology, politics and society [Bell 1974: 192; Schon 1967: 200]. The question is: how do we respond to this new situation? One of the answers must lie in the kinds of choices we make and the kinds of values that underlie these choices [Mumford and Sackman 1975: 325].

The evidence from this study shows that in the four organizations that were investigated, values did influence objectives, design and assimilation of computer systems.

In the government department, Asbestos Ltd., the international bank and, to a lesser extent, Chemco, the introduction of a new level of technology was recognized as an opportunity to increase job satisfaction and this was done for humanistic as well as practical reasons. Asbestos Ltd. and the international bank not only made job satisfaction a systems objective, they gave the clerks themselves a major role in the design of the new work organization which would be associated with the computer system. The government department was very aware of the need to provide job satisfaction for its clerical officers; Chemco of the need to do this for user management.

Schon believes that there are two powerful pressures in society today; these are the call to "return" and the call to "revolt". The first tells us to return to the old ways and standards, to try and retrieve the vanished stability. The call to revolt says that there is increasing alienation in society and that we must develop new values and objectives [Schon 1967: 202]. The practical problem for those who, like the author, believe that technology can be an instrument for human well-being rather than a threat to this, is how to assist the development of these new values. Sociologists have always been excellent at diagnosis after the event but what is required today is "real time" sociology, a speed in problem solution sufficient to influence the events to which the problem applies.

In our four organizations we have seen that the values of top management were an important influence on the consequences of change and as Fox suggests [1974: 59], in the future we must expect top management to show a "keener alertness to the needs and aspirations of those upon whose identification, commitment and loyalty, they are heavily dependent". In Asbestos Ltd. the influence of a group of technical systems designers who felt strongly that their computer systems should assist an increase in job satisfaction enabled this result to be achieved. This study has provided no evidence to support the belief that computer technologists are influenced by a powerful technical ethic in which technical gains are striven for at all costs. The reality in the four organizations was that technologists and managers shared a common value system that this was either purely rational or both rational and humanistic.

Vickers has pointed out the responsibility of the policy maker to strive for what he believes to be right and this was a necessary component of change in the four organizations [1973: 178]. Even though some clerks were not satisfied with their jobs, the demand for change did not come from them and in the organizations which made job satisfaction a goal their values tended to be more conservative than those of the managers and the systems designers. Fox points out that employees do not question everything, certain things are legitimized and up to now these have included work roles [ibid: 91]. Similarly, union values have made them demand higher pay or more job security for their members, but not more interesting work, although these values are now starting to change.

There are many suggestions on how a change of values is and can be made. Kuhn has put forward the scientific "crisis" theory, which suggests that when existing paradigms cease to solve problems, they will be replaced with new ones [Kuhn 1962: 77]. The strategy here would be to draw attention to this kind of failure. Parsons believes that a system's inherent tendency to adjust when discrepancies occur between its needs and the external environment will keep it in a state of stability, providing there is value consensus on what these needs are [Parsons 1964: 21; 1964: 36]. This suggests that an emphasis should be placed on the development of shared values of a kind different from those of the past. Hedberg maintains that an increasing awareness of change of consequences can be achieved through making values explicit [1965: 10]. Fox's view is that low trust relationships between management and employees should be replaced by high trust relationships, and that providing employees with opportunities for the use of discretion is a manifestation of this trust [1974: 114].

All of these diagnoses and suggestions are insightful and helpful and will be increasingly debated in the future. In this book the emphasis has been on the role of computers as a vehicle for change and the author's belief is that this technology can be used as an agent for "technological bonding", so that it contributes to an enhancement of the quality of working life. Computers and automation can be an agent for good or bad, the quotations below are from writers who appreciate its potential as an agent for good.

> Automation is not gadgeteering, it is not even engineering; it is a concept of the structure and order of economic life, the design of its basic patterns integrated into a harmonious, balanced and organic whole [Drucker 1962: 229]

> Humanistic automation is the elevation of human intelligence and the enhancement of social well-being through the use of computers for the experimentally regulated evolution of social systems in response to changing human needs. [Sackman 1967: 211]

12. The Management of Future Technology: A Problem of Uncertain Visions and Values

This book has considered how values affect the design of technology and work systems today. The important question is what values will influence the way we use technology in the future. At present we see technology as today's hope but tomorrow's bogey man. Christopher Evans, computer scientist and psychologist, has described our situation as a three-stage revolution which has already begun. Stage one (1975-1982) he sees as the dawn of public awareness in terms of computers, the stage of gimmicks and toys which is already evident. Stage two (1983-1990) will be when we see the first large-scale changes in the patterns of work and education, economics and social life. Stage three (1991-2000) will be signalled by political upheavals, and the vastly important emergence of the ultra-intelligent machine [Evans 1979]. If we are already in a revolutionary mode, a revolution being defined as major changes in roles, social structures, norms and attitudes, then we have a great many things to do. We need to gain an understanding of the societal implications of the new technology, and of the different kinds of future it can lead us to. We need also to develop strategies for its control and management so that we are able to guide technology in the direction we want it to take. The author's namesake, Lewis Mumford, once said: "Science discovers, technology executes, man conforms", implying that we follow blindly wherever technology takes us. Can we afford to do this in the future or must we, as members of society, assume a more interventionalist role than we have perhaps been willing or able to do in the past?

The subject of technology and its impact on society is a deeply philosophical issue which cannot be understood by examining technology alone. The author has incorporated into the title of this chapter what she believes are the three most important concepts which any debate must cover. She sees these as *visions* of a desired future, created through generally approved humanistic *values* and secured by a skillful *management* of change.

Although we are concerned with philosophy, we are also concerned with application and practice. The discussion must always be at three levels. What can technology offer us, what do we want from it and how can we achieve what we want? Stating the problem in this way immediately draws our attention to a fundamental difficulty. Who is "we"? Unfortunately "we" is not a closely knit group of people, all with similar interests and objectives and with common values and visions. "We" in contrast, is a complex network of vested interests, many of these interests in conflict with each other. Even if visions of a desired and desirable future have elements in common, each group wishes to take a different route to its utopian society. Our management problem then becomes very complex. It is not a question of leading people towards a commonly desired promised land but one of identifying and reconciling the interests and objectives of a number of powerful groups, each one of which is going to be affected by the form and pace of new technology.

This book has been concerned with the use of technology in the office and it is the office that is seen by forecasters as being most affected by technology in the future. Some see its impact as dramatic with the new automated methods of handling automation decimating office employment. Let us briefly consider the nature of the technology that will affect the office in the next five years.

12.1. THE NATURE OF NEW OFFICE TECHNOLOGY

New technology in the office is associated with the automatic transfer and manipulation of textual information. Whereas traditional data processing handled numerical information, information technology deals with words. We have all become familiar with the term ''microprocessor'' but it should be noted that the microprocessor is only an electronic component. To use it, it has to be combined into a system with other components. These consist of devices such as key-boards, optical display screens, printers and programmed instructions (software). The importance of the microprocessor to new technology is that it can be very cheaply mass produced. It therefore provides the possibility of bringing computer aids to all of us. Word processing systems are the first step towards the electronic office. The eventual electronic transfer and storage of information (electronic mail and filing) within and between companies has the potential to replace hard copy memos, letters and files.

These developments are seen as producing a fundamental reorganization of business communications, secretarial services and the offices in which these take place. Word processing marks the beginning of this change. The technology is at present in a rapid state of development and becoming dramatically cheaper.

12.2. THE AUTOMATED OFFICE

There are two possible scenarios here: (1) the growth of a computer-based information system with linked terminals and a central filing system is one concept; (2) distributed systems based on decentralized information is another.

Paper as a vehicle for storing information is likely to go and electronic mail will become commercially attractive – at least within companies. The use of a secretary as an intermediary with the machine may become unnecessary. Managers can input their information directly – by simplified keyboards or voice.

The lives of office workers and their managers are likely to change dramatically with women being particularly threatened, as office work is so much a feminine occupation. There are over one million typists and secretaries in the UK.

We can hypothesize that it is unlikely that this group or their managers will ac-

280

cept this change easily unless they believe that it is in their interests to do so and that personal gains will result.

12.3. THE QUESTIONS THAT NEED ASKING

Given the size of the potential problem it is important to ensure that significant questions are asked about it. I would see these as being:

At a societal level
—*How* do we *control* this new technology in the best interests of society as a whole?
—Can we solve the future problems within the present framework of society or will a basic reorganization of society be required?
—Can we initiate, progress and institutionalize new and desired forms of relationship between man and his environment, particularly his work environment?

At an organizational level
—Who *owns* the new technology (whose work system is it?)?
—Does it belong to top management, to the engineers and computer specialists who designed it or to the clerks and their managers who use it?
—Who should have responsibility for *designing* the new technology and its organizational context? Traditionally, this has been left to the technologists. Should it not now also involve the user groups?
—*Who* should *control* the new technology within the organization? Is it top management, who decide to introduce it and who pay for it, the technologists who design and understand its engineering or the users who operate it?
—Who should decide to *abandon* it if its use is proving irreconcilable with human needs and a humanistic work environment?

A further and even more fundamental question must now be raised. Are there even rational logical answers to these questions? Some experts would argue that the answer is "no". Professor Stafford Beer has said: "It might be more helpful to talk about tribal customs than about logical analysis, about myths and taboos rather than understanding and intellection, about conversion rather than conviction".[1]

Ward [1973: 114] has said succinctly that "technology is a tyrant's tool". He has also said when personal interests are threatened, objectivity requires the qualities of saints. Very often those most affected are asked to trade personal benefits for long-term gains that others will receive.

Others have criticized the technologists themselves, suggesting that they have little social responsibility. Tom Lehrer once sang: "So long as they go up, who cares where they come down, that's not my department said Werner Von Braun".

---

1. Presidential address to Cybernetics Conference, September 1979; quoted in [Ward 1973].

We cannot escape the debate or the responsibility so easily, however. The introduction and use of technology is a management problem in which all groups in society have an interest and it is beholden on all of us to try and think clearly and logically about the management process. We have to take decisions on the following critical issues:

—The kinds of objectives that are set for the new technology.
—The design of the community and organizational systems that surround the new technology.
—How these systems shall be implemented.
—The implications of these decisions for society as a whole and for particular organizations and groups within society.

These decisions are unlikely to be left solely with any one group. They will have to be shared amongst the trade unions, the office workers and their managers, the technologists, the community and the government. All groups with very different values, attitudes and objectives.

12.4. STRATEGIES FOR CHANGE

Successful change involves the *setting and attainment of goals and objectives, adaptation,* so that an organization and its members can move easily from state A to state B; *integration* – the process of fitting new technology into an appropriate organizational framework so that human, technical and business needs are met at one and the same time, and the interaction of technology, task structures, attitudes and behaviour to produce a high quality work environment and a satisfied, skilled and motivated work force. Once the change has been introduced and become operational then the new system must be maintained in a stable state until it is necessary to change it once again.

An important question is what strategies are available to assist these change processes? Mary Parker Follet [1924] has suggested that there are three principal strategies for change. These are *domination, compromise* and *integration* – using integration in the sense of meeting a variety of interests and objectives at one and the same time. *Domination* she sees as an undesirable strategy as it is a victory of one side over the other. *Compromise* is also unsatisfactory because neither side gets what it wants. *Integration* in contrast, means finding a way that will include both what A wants and what B wants, and also what C and D want, with neither side having to sacrifice anything. In her view, this is the way of progress. Looking for high level solutions that can meet the needs of many groups without compromises will not be easy, but perhaps we should try to achieve this.

12.5. THE NEEDS AND INTERESTS OF DIFFERENT GROUPS

What are the needs and interests of the different groups that will be affected by the new technology of the offices of the future? Starting at the bottom of the organization, there are the white collar workers. They see themselves particularly threatened by the new technology and resistance to it is growing in many European countries. Their fear is of course that of unemployment. They see themselves displaced by small chips of silicon which are very cheap to make and which have the power of what a few years ago was a very large computer. At the moment the picture of what is going to happen to employment is very uncertain and there are conflicting forecasts. For example, the consultant firm Arthur D. Little has produced a study of the years 1978-1985 and forecast that the net effect of the new technology on employment in Western Europe and the USA will be an increase of *one million* jobs. Almost at the same time as the Arthur D. Little report appeared, Apex, a British white collar trade union, forecast that by 1980 the UK would have lost 25,000 jobs. Their head of research has been quoted as saying: "A typist in central London probably costs £6,000 to £7,000 per annum in salary, employment costs, social insurance, etcetera for an employer . . . . You can replace a typist by installing a word processor which costs £4,000". Most trade unions appear pessimistic about the impact of new technology on employment and Clive Jenkins in his recent book *The Collapse of Work* suggests that we must abandon the "work ethic". Perhaps this is already starting to happen. Today, in Britain, we regard two million unemployed as acceptable, a figure that would have been regarded with disbelief and horror ten years ago.

What is the managers' relationship with new technology? It is managers that have primary responsibility for its introduction and they are going to require great skill to do this. Clive Jenkins says that "management will be under greater pressure than ever before" and it is clear that the management problems of the years ahead will be tremendous.

There are also the engineers and technologists who design the new technology. They have a vested interest in change and progress and in working at the frontiers of knowledge. They are experts in rationality but in the future we must ask them to be experts in humanity too. Science promises men power, but as so often happens when people are seduced by power, the price paid can be servitude and impotence [Weizenbaum 1976]. Professor Singer in a book called *In Search of a Way of Life* has said that for humanity to grow in power "it is necessary for every mortal man to spend every passing day in unsparing effort to advance humanity along each of the three dimensions of progress – science, morals, art".

What about the role of government in this revolutionary change? There is likely to be a demand for proper mechanisms of social control over technologies which have a social impact and this includes nuclear power and biotechnology as well as information technology. Governments will be required to regulate and monitor the new technology and to reduce its possible adverse effects. However, a problem here is that information technology is rapidly becoming a transnational industry with a few manufacturers controlling a world-wide industry. This makes national control very difficult.

12.6. CONCLUSION

We are moving into the third industrial revolution, a revolution which may be faster and more dramatic than the previous two. How can this change be managed so that we reap major benefits without at the same time incurring major losses? At a national and organizational level perhaps the solution offered by Mary Parker Follet is the best one. All groups with an investment in the consequences of the change – workers, managers and technologists – must come together to find a strategy for change that furthers all interests and leads to progress for all. The new technology must not lead to winners and losers or exploiters and exploited. Such a strategy requires some shared values and common interests, perhaps related to a vision of future society that all wish to attain. The author believes that a contribution to an integrated and agreed solution must come from increased participation in decisions. Her own activity in industry and commerce with groups affected by computer systems has demonstrated that they can be helped to acquire the skills to design their own job structures around new technology. In this way clerks can diagnose their efficiency and job satisfaction needs and work with the technical specialists to ensure that the design of technology and its organizational context is directed to meeting these needs as well as wider economic objectives. But without values and methods to achieve some consensus on the route technical change should take, the next few years could easily become a battle-ground between managers and workers.

We always have a choice on the route to take; we need to ensure that our decisions on how to use technology in the future are based on humanistic values, an agreed vision of a socially desirable society and the knowledge that we have the skills to successfully manage revolutionary technical change so as to achieve happiness, dignity and freedom for all.

Let us end with the wisdom of Stafford Beer. He says:

> The future is something we use our freedom to determine, rather than something that is lurking out there, and will happen to us, unless we are mighty smart. We can make, rather than prophesy, the future. [1974: 91]

Appendix A. The influence of values on strategies

Organization and group	Value category and expression	Value (Parsons)	Change variable	Manifest consequences	Latent consequences	Effect on change variable
Chemco home sales On-line system Systems designers and management	**Economic** System should produce financial gains through improved cash flow	Cognitive	Goal attainment	Computer is used to increase speed of order processing/faster payment of bills	Clerks approve of this increase in efficiency job satisfaction is increased	This goal is satisfactorily attained
Chemco home & export sales On-line systems Systems designers	**Technical** Need to gain experience in design of on-line computer systems	Cognitive	Goal attainment	On-line system designed for users who welcomed this development	Users see systems as improvements on previous batch systems; batch system was liked in home sales, disliked in export	This goal is satisfactorily attained
Chemco home and export sales Batch and on-line systems Systems designers and management	**Rational** Process orders more efficiently; obtain more accurate information	Cognitive	Goal attainment	System designed that is similar to manual system in concept	*Functional in home sales* Clerks approve of this increase in efficiency *Dysfunctional in export* System is not seen as helping the efficiency of complex export work, clerks are frustrated	This goal is satisfactorily attained in home sales, not in export
Asbestos Ltd. On-line system Systems designers	**Socio-technical** To remove the barrier between clerks and computer that existed with the batch system; design system to meet the needs of the clerks	Cognitive Cathectic	Goal attainment	Introduction of on-line system with VDUs, enquiry facility and new work organization	Change in attitude from dislike of batch system to enthusiasm for on-line system	This goal is satisfactorily attained
International bank On-line system Management and systems designers	**Socio-technical** To improve the job satisfaction of clerks through reorganizing work; to improve efficiency in foreign exchange department	Cathectic Cognitive	Goal attainment	New form of work organization based on currency divisions associated with introduction of on-line system which eliminated much routine work	More efficiency, more work variety, but grading and staffing problems lowered morale	This goal is partially attained

(continued on p. 286)

285

(Appendix A. The influence of values on strategies, continued from p. 285)

Organization and group	Value category and expression	Value (Parsons)	Change variable	Manifest consequences	Latent consequences	Effect on change variable
Government department (Inland Revenue) Systems designers and management	**Social** Increase job satisfaction	**Cathectic**	Goal attainment	Routine work is put onto the computer, complex work is given to tax officers	Some stress from constant problem solving	This goal is satisfactorily attained
Government department (Inland Revenue) Management	**Rational** Increase efficiency, use computer potential	**Cognitive**	Goal	Centralized Inland Revenue Centre is built	Difficulties in getting staff; loss of face-to-face relationships with the public	Goal attained with some social costs
Chemco home & export sales Batch and on-line systems Management and systems designers	**Rational/adaptive** To secure user interest in and commitment to the system	**Cognitive Cathectic**	Adaptation	User management very much involved in design; committees set up to involve clerks, but did not last in export	*Functional in home sales* Management understood system, clerks liked system *Dysfunctional in export* Some, but not all clerks adapted to system	Good adaptation at management level in home and export; good adaptation at clerical level in home, not in export
Asbestos Ltd. On-line system Systems designers	**Social/adaptive** Desire to use knowledge of clerks and secure their commitment	**Cathectic Cognitive**	Adaptation	Consultative participation structure set up; clerks assume responsibility for some design tasks	Learning process takes place; clerks become knowledgeable about their own and the department's needs	Good adaptation at clerical level; this assisted management adaptation
International bank On-line system Management and systems designers	**Social/adaptive** Desire to give clerks responsibility for improving their own job satisfaction	**Cathectic**	Adaptation	Representative participation structure set up; clerks assume responsibility for all social system design tasks	Learning process takes place; clerks become expert social system designers	Good adaptation at clerical level; system is easy to implement
Government department (Inland Revenue) Management	**Rational** Ensure staff are competent to operate computer based system	**Cognitive**	Adaptation	Comprehensive training programmes for new work and organization	Some adaptation problems with older tax officers	Adaptation to operational aspects of new system

Organization	Orientation	Type	Level	Action	Effect	Outcome
Chemco home and export sales / Batch and on-line systems / Management and systems designers	**Rational/integrative** / Belief that clerks will adjust to new technology and make it work	**Cognitive**	Integrative	All clerks operate VDUs; adjustments made to work organization as clerks ask for these	*Functional in home sales* / Clerks feel they have control and skill; dept. has clerical flexibility / *Dysfunctional in export* / Problem pushed from one group of clerks to another as each group complains	Good socio-technical integration in home sales; good integration with customer environment; less good integration in export; resistance to change
Asbestos Ltd. / On-line system / **Systems designers,** management, and clerks	**Social/integrative** / Willingness to let clerks **choose the system which** best meets their needs	**Cathectic** / **Moral** / **Cognitive**	Integrative	Dept. organized to auto**nomous, multi-skilled** groups; well integrated jobs	Improved job satisfaction and **efficiency; good customer** service; commitment to technology; less routine/ stress	Good socio-technical integra**tion; good integration** with customer environment; personal problems
International bank / On-line system / Management, clerks, systems designers	**Social/integrative** / Willingness to let the clerks design the system which best meets their needs	**Cathectic** / **Moral** / **Cognitive**	Integrative	Dept. organized into autonomous groups with potential to become multi-skilled	Good customer service; less routine work	Good socio-technical integration; good integration with customer environment
Government department (Inland Revenue) / Systems designers and management	**Social/integrative** / Design effective socio-technical system	**Cathectic** / **Cognitive**	Integration	System works well within I.R. boundary	Reduced social interaction between tax officers and public is regretted by both sides	Poor organization/environment integration from tax officers' viewpoint
All four organizations	**Social/rational** / Desire to keep system in state of socio-technical equilibrium	**Cathectic** / **Moral** / **Cognitive**	Pattern maintenance	Monitor stability of social and technical systems; make adjustments when required	Maintenance of job satisfaction and efficiency	This may be difficult to achieve in Chemco export sales and international bank

287

Appendix B.
Job Satisfaction Questionnaire

This questionnaire was used to measure job satisfaction. In the international bank and Chemco the questionnaire was used both before and after the implementation of the new computer system. The pre-change version did not ask staff to compare the batch with the on-line system, otherwise it was the same.

This is a follow up of the first questionnaire you were kind enough to complete. Its aim is to find out how the new on-line system affects your work. You will remember that I am trying to obtain information which will enable jobs to be designed in such a way that people really enjoy doing them. I shall be most grateful if you will help me once again.

There are no right or wrong answers. The best answer is your personal opinion.

The study is being carried out by myself, Enid Mumford of Manchester University.

Each questionnaire is completely anonymous, the only information you are asked to give is your job title, sex and age. This helps the researcher to find out if particular jobs have special problems and if people of different ages, or men and women, have different points of view. However, if you would rather not give this information then leave this page blank.

No-one will see the questionnaires except myself. I will collect them from you personally and immediately remove them to Manchester University.

I would have liked to talk with you all individually but in order to get information from as many people as possible in as many departments as possible a "fill in yourself" questionnaire seems the quickest way. Most of the questions only require a tick but some ask for additional information which will assist me in understanding what you like and dislike about your job. If you would care to write any other comments about your work on the questionnaire I should welcome this.

Please ask me if you have any problems with any of the questions.

To begin, please fill in the following

> Job title . . . .
> Age . . . .
> Sex (put M or F) . . . .

How often during your workday is your work fairly routine and how often is doing the job well a real challenge to your ability or ingenuity (please tick)?

My work is:

1. ..... Almost always a challenge to my ability or ingenuity
2. ..... Usually challenging
3. ..... About half routine and half challenging
4. ..... Usually routine
5. ..... Almost always routine

If you have ticked 1, 2 or 3 is it *challenging* because (please tick)

	You have difficult problems to solve
.....	You have a number of very different kinds of tasks to perform
.....	You have a great deal of responsibility
.....	The work requires special skills and knowledge

or If it is challenging for any other reason, please write the reason below.

....
....
....

Is the amount of challenge in your job

..... Too much
..... About right
..... Too little

If you have ticked 3, 4 or 5 is it *routine* because (please tick)

	There are very few problems to solve
.....	You do the same set of tasks all day and every day
.....	You have little responsibility
.....	The work requires little skill and knowledge

If it is routine for any other reason, please write the reason below.

....
....
....

Is the amount of routine in your job

..... Too much
..... About right
..... Too little

How often do you get a sense of achievement from your work? (please tick)

..... Almost every day
..... About once a week
..... Once every few weeks
..... About once a month
..... Hardly ever

What gives you the most sense of achievement in your work? ....

Would you like to get a sense of achievement

..... More often
..... About the same

Comparing the new on-line computer system with the old batch system, would you say that the amount of challenge in your job is (please tick)

..... Greater than before
..... About the same as before
..... Less than before

Would you say that you get a sense of achievement from your work

     . . . . . More frequently
     . . . . About the same
     . . . . Less frequently

If you have said "greater" or "more frequently" above, could you say why this is the case?
. . . .
. . . .

On most days on your job, how often does time seem to drag for you? (please tick)

     . . . . . About half a day or more
     . . . . . About one-third of the day
     . . . . . About one-quarter of the day
     . . . . . About one-eighth of the day
     . . . . . Time never seems to drag

What causes time to drag most of all? . . . .
. . . .

Comparing the new on-line computer system with the old batch system, does time drag for you

     . . . . . More often
     . . . . . About the same
     . . . . . Less often

Some people are completely involved in their job – they are absorbed in it night and day. How involved do you feel in your job? (please tick)

     . . . . . Very little involved; my other interests are more absorbing
     . . . . . Slightly involved
     . . . . . Moderately involved; my job and my other interests are equally absorbing to me
     . . . . . Strongly involved
     . . . . . Very strongly involved; my work is the most absorbing interest in my life

When there is a time limit or deadline or target date for your work, how "tight" are these deadlines usually? (please tick)

     . . . . . Much more than enough time to do the job
     . . . . . Ample time
     . . . . . Just barely enough time
     . . . . . Too little time to do the job

Has the on-line computr system made deadlines

     . . . . . Easier to meet
     . . . . . Made no difference
     . . . . . Harder to meet

If you have ticked "easier to meet" why has it had this effect? . . . .
. . . .

About what per cent of your time on the job are you working with a time limit or deadline or target date for the work you are doing? (please tick)

There is a time limit or deadline or target date for my work:

..... 90 per cent or more of the time
..... about 75 per cent of the time
..... about 50 per cent of the time
..... about 25 per cent of the time
..... about 10 per cent of the time
..... never

When deadlines or time limits are being set for work you are doing, which of the following appears to be the *most important* influence on how these limits are set (TVA). (Put a number 1 next to the most important influence)

..... Requirements of *people outside* your own section or department
..... Your *own estimate* of how long the job will take
..... Your *supervisor's estimate* of how long the job will take

Now go back and put a number 2 next to the second most important influence on what the time limits or deadlines are. Put a number 3 next to the third most important influence.

Would you prefer to have

..... more targets and deadlines than now
..... about the same as now
..... fewer than now

If you make a largish error in your work on average how soon would it be brought to your attention? (please tick)

..... Almost immediately
..... The same day
..... Within two days
..... Within a week
..... Within a month
..... Longer than this

Who would bring it to your attention? (please tick)

..... I would spot it myself
..... Another colleague
..... A supervisor or manager in this department
..... A supervisor or manager in another department
..... A client or customer

Would you like to learn about errors (please tick)

..... More quickly than at present
..... As now
..... More slowly than at present

Who would you prefer to hear about them from? ....

What kind of errors cause you most problems? . . . .
. . . .

Comparing the new on-line computer system with the old batch system, do errors

       . . . . . Show up more quickly now
       . . . . . There is no difference
       . . . . . Take longer to show up

If errors show up more quickly, why is this? . . . .
. . . .

When you get a job to do, how often is it completely up to you to decide how to go about doing it? (TVA) (please tick)

       . . . . . Hardly ever
       . . . . . About one-quarter of the time
       . . . . . About half of the time
       . . . . . About three-quarters of the time
       . . . . . Almost always

When you finish a given piece of work, how often do you have any say at all about what work you will be doing next? (please tick)

       . . . . . Almost always
       . . . . . About three-quarters of the time
       . . . . . About half of the time
       . . . . . About one-quarter of the time
       . . . . . Extremely rarely or never

When you have several things to do on the job, how often is it up to you to decide which you will do first and how often are you expected to do things in a set order? (TVA) (please tick)

       . . . . . Almost always expected to do things in a set order
       . . . . . Sometimes have a set order, sometimes not
       . . . . . Almost always up to me to decide which I'll do first

If you could choose, would you like (please tick)

More freedom to decide on the best way to to a par-	More choice in what you do	More freedom in choosing the order of
. . . . . ticular job	. . . . . next	. . . . . of work
. . . . . As now	. . . . . As now	. . . . . As now
More guidance and	More guidance and	Prefer to have a stricter order
. . . . . less freddom	. . . . . less choice	. . . . . of work

Would you say that your job provides you with an opportunity to make decisions and use your own judgement? (please tick)

       . . . . . Yes, to a considerable extent
       . . . . . Yes, to some extent
       . . . . . No, hardly at all

292

If you ticked yes, what kinds of decisions do you most enjoy taking? (please give examples below)

. . . .
. . . .
. . . .

Would you like to have (please tick)

. . . . . The opportunity to take more decisions and use more judgement
. . . . . The same as now
. . . . . Fewer decisions than now

Comparing the new on-line system with the old batch system, do you

. . . . . Take more decisions than before
. . . . . There is no difference
. . . . . Take fewer decisions than before

If other people you have contact with on the job don't do their work correctly or on time, how often would this create problems for your own work? (TVA) (please tick)

If this happened it would create problems for my work

. . . . . Almost always
. . . . . Usually
. . . . . About half the time
. . . . . Occasionally
. . . . . Very rarely or never

If you don't do your own job correctly, or fast enough, how often would this create problems for someone you have contact with on the job? (TVA) (please tick)

If this happened it would create problems for someone I have contact with

. . . . . Almost always
. . . . . Usually
. . . . . About half the time
. . . . . Occasionally
. . . . . Very rarely or never

How many people with whom you have contact on the job could create problems for your work if they didn't do *their* work correctly or on time? (TVA) (please tick)

. . . . . None of them could create problems for me
. . . . . Two people could create problems for me
. . . . . Three people could create problems for me
. . . . . Four people could create problems for me
. . . . . Five to ten people could create problems for me
. . . . . More than ten could create problems for me

If you didn't do your work correctly or on time, for *how many* people with whom you have contact on the job would this create problems? (please tick)

..... None of them would have problems
..... Two people would have problems
..... Three people would have problems
..... Four people would have problems
..... Five to ten people would have problems
..... More than ten people would have problems

Would you prefer to have the kind of job where you are less dependent on the work of others or others are less dependent on your work? (please tick)

..... Yes
..... No
..... Makes no difference to me

Comparing the new on-line computer system with the old batch system, are you now

..... Less dependent on the work of others
..... There is no difference
..... More dependent on the work of others

In connection with your job how much chance do you get to see a piece of work right through from start to finish, without having to pass it on to someone else?

..... Very little or no chance
..... Litte chance
..... Some chance
..... A good chance
..... An excellent chance

If there is the possibility of seeing a piece of work right through from start to finish, how often are you able to do this?

..... Once a day
..... About once a week
..... About once a month
..... Less frequently than this

If you could choose, would you like your job (please tick)

To be large enough for you to be able to see a piece
..... of work through from start to finish
..... To be as now
To be less extensive than now so that tasks were
..... smaller in size

Do you always understand exactly what you have to do and why you have to do it?

..... Always understand
..... Understand about 75 per cent of the time
..... Understand around 50 per cent of the time
..... Understand around 25 per cent of the time
..... Hardly ever understand

294

Do you always agree with what you have to do and why you have to do this?

..... Always agree
..... Mostly agree
..... Hardly ever agree

If you have ticked "hardly ever agree", could you write your reasons for not agreeing below? ....
....

How much of your work is checked, inspected or reviewed by someone else?

..... 100 per cent of my work is checked, inspected or reviewed
..... About 75 per cent or more of my work is checked, inspected or reviewed
..... About 50 per cent of my work is checked, inspected or reviewed
..... About 25 per cent of my work is checked, inspected or reviewed
..... Almost none of my work is checked, inspected or reviewed

Would you like to have (please tick)

..... Less checking than now
..... About the same
..... More checking

Comparing the new on-line computer system with the old batch system, is:

..... Less of the work now checked by someone else
..... There is no difference
..... More of your work now checked by someone else

In your job how often do you find yourself unable to get information or materials needed to carry out your job properly?

..... Several times a week
..... About once a week
..... Several times a month
..... About once a month
..... Once every few months
..... Never

Comparing the new on-line computer system with the old batch system, do you

..... Get more information now
..... There is no difference
..... Get less information now

In some jobs there are detailed rules about what is the right way to do the job; the correct procedures to adopt. How is it with your job? (please tick)

..... Almost everything is covered by rules or set procedures
..... Most things are covered by rules or set procedures
..... About half and half are covered by rules or set procedures
..... There are a few rules and set procedures
..... There are hardly any rules or set procedures

Taking your job as a whole, would you say that you enjoy your work, or not?

. . . . . Enjoy it very much
. . . . . Quite enjoy it
. . . . . Do not enjoy it

If you have ticked "enjoy it very much" or "quite enjoy it" please indicate below how often you enjoy it?

. . . . . Enjoy it all the time
. . . . . Enjoy it 75 per cent of the time
. . . . . Enjoy it 50 per cent of the time
. . . . . Enjoy it 25 per cent of the time

Comparing the new on-line computer system with the old batch system, do you

. . . . . Enjoy work more now
. . . . . There is no difference
. . . . . Enjoy work less now

Can you make any suggestions as to how your job could be made more efficient? (please write below)

Can you make any suggestions as to how your job could be made more enjoyable or satisfying for you? (please write below)

Think of a job you would regard as ideal from the *work* point of view. How does your present job compare with this? (please tick on the scale below as follows: if your present job is very close to your ideal job then tick 6, if it is very far away tick 1 and so on)

Ideal job 7
6 . . . . .
5 . . . . .
4 . . . . .
3 . . . . .
2 . . . . .
1 . . . . .

Do you find that the use of a computer by this department (please tick)

. . . . . Increases your efficiency
. . . . . Makes no difference
. . . . . Decreases your efficiency

If the computer increases or decreases your efficiency, how does it do this?

296

Do you find the use of a computer by this department (please tick)

..... Increases the interest of your work
..... Makes no difference
..... Decreases the interest of your work

If the computer increases or decreases the interest of your work, how does it do this?

Set out below are pairs of descriptive comments – each of the two comments representing opposite positions. To fill in you tick on the part of the scale which coincides with your own point of view. Thus, if you think the left hand comment is exactly right you would place a tick on the black point under 1. If you think that neither comment is correct and the real position is somewhere between the two you would tick the black point under 4. If you think the right hand comment is correct, or nearly correct, you would tick under 7 or 6.

[scales are reproduced on the four following pages]

Please tick where you think this firm, bank, civil service department fits on the scale below

It has a strong belief in the importance of shared values – that employees should agree with its objectives and the way it carries out its operations

1 2 3 4 5 6 7

├─┼─┼─┼─┼─┼─┤

A multiplicity of different values exists and is tolerated here. It is quite acceptable for employees to hold very different views on what the firm should be doing and how it should be doing it

It likes to have disciplined employees who will put the firm's/office's interests first and willingly accept orders and instructions

├─┼─┼─┼─┼─┼─┤

It does not believe in very strict discipline and tries to provide a situation in which people can pursue their own interests

It prefers to use standardized methods and procedures whenever possible as it believes that this assists efficiency in this kind of firm/office

├─┼─┼─┼─┼─┼─┤

It likes employees to work out their own methods for doing things and to use their own judgement when taking decisions

It places a great deal of emphasis on efficiency and high production and less on personal qualities such as friendliness, trustworthiness, cooperation etc.

├─┼─┼─┼─┼─┼─┤

It places a great deal of emphasis on personal qualities such as friendliness, trustworthiness, cooperation etc. and less on efficiency and high production

It feels the need to organize work activities into tightly structured jobs which are clearly defined and do not permit a great deal of individual discretion

├─┼─┼─┼─┼─┼─┤

It tries to organize work in such a way that employees have loosely defined and structured jobs which permit a great deal of individual discretion

298

Please tick where you would like this firm, bank, civil service department to fit on the scale below.

	1 2 3 4 5 6 7	
It should have a strong belief in the importance of shared values – that employees should agree with its objectives and the way it carries out its operations	├──┼──┼──┼──┼──┼──┤	A multiplicity of different values should exist and be tolerated here. It should be quite acceptable for employees to hold very different views on what the firm should be doing and how it should be doing it
It should try to create disciplined employees who will put the firm's/office's interests first and willingly accept orders and instructions	├──┼──┼──┼──┼──┼──┤	It should have an easy-going kind of discipline and try to provide a situation in which people can pursue their own interests
It should try to use standardized methods and procedures whenever possible as this would assist efficiency in this kind of firm/office		It should encourage employees to work out their own methods for doing things and use their own judgement when taking decisions
It should place a great deal of emphasis on efficiency and high production and less on personal qualities such as friendliness, trustworthiness, cooperation etc.		It should place a great deal of emphasis on personal qualities such as friendliness, trustworthiness, cooperation etc. and less on efficiency and high production
It should try to organize work activities into tightly structured jobs which are clearly defined and do not permit a great deal of individual discretion	├──┼──┼──┼──┼──┼──┤	It should try to organize in such a way that employees have loosely defined and structured jobs which permit a great deal of individual discretion

Please tick on the scale below what you believe to be the best form of department structure for clerical staff in general.

	1 2 3 4 5 6 7	
Jobs should be clearly defined, structured and stable	├──┼──┼──┼──┼──┼──┤	Jobs should be flexible and permit group problem solving
There should be a clear hierarchy of authority with the man at the top carrying ultimate responsibility for all aspects of work	├──┼──┼──┼──┼──┼──┤	There should be a delegation of authority and responsibility to those doing the job regardless of formal title and status
The most important motivators should be financial, e.g., high earnings and cash bonuses	├──┼──┼──┼──┼──┼──┤	The most important motivators should be non-financial, e.g., work challenge, opportunity for team work
Jobs should be carefully defined by O&M department, management services or supervision and adhered to	├──┼──┼──┼──┼──┼──┤	The development of job methods should be left to the group and individual doing the job
Targets should be set by supervision and monitored by supervision	├──┼──┼──┼──┼──┼──┤	Targets should be left to the employee groups to set and monitor
Groups and individuals should be given the specific information they need to do the job but no more	├──┼──┼──┼──┼──┼──┤	Everyone should have access to *all* information which they regard as relevant to their work
Decisions on what is to be done and how it is to be done should be left entirely to management	├──┼──┼──┼──┼──┼──┤	Decisions should be arrived at through group discussions involving all employees
There should be close supervision, tight controls and well maintained discipline	├──┼──┼──┼──┼──┼──┤	There should be loose supervision, few controls and a reliance on employee self-discipline

Please tick on the scale below how you see the general characteristics of the majority of staff in this department.

	1	2	3	4	5	6	7	

They
Work best on simple, routine work that makes few demands of them
├─┼─┼─┼─┼─┼─┤
They
Respond well to varied, challenging work requiring knowledge and skill

They are
Not too concerned about having social contact at work
├─┼─┼─┼─┼─┼─┤
Regard opportunities for social contact at work as important

They
Work best if time and quality targets are set for them by supervision
├─┼─┼─┼─┼─┼─┤
Are able to set their own time and quality targets

Work best if their output and quality standards are clearly monitored by supervision
├─┼─┼─┼─┼─┼─┤
Could be given complete control over outputs and quality standards

Like to be told what to do next and how to do it
├─┼─┼─┼─┼─┼─┤
Can organize the sequence of their work and choose the best methods themselves

Do not want to use a great deal of initiative or take decisions
├─┼─┼─┼─┼─┼─┤
Like, and are competent to use initiative and take decisions

Work best on jobs with a short task cycle
├─┼─┼─┼─┼─┼─┤
Able to carry out complex jobs which have a long time span between start and finish

References

Ackoff, R.L. and P. Rivett (1963), *A Manager's Guide to Operational Research*. New York: Wiley.

Allport, G.W. (1935), "Attitudes", in C.M. Murchinson (ed.), *Handbook of Social Psychology*. Worcester, Mass.: Clark University Press.

Allport, G. (1954), *The Nature of Prejudice*. New York: Doubleday.

Argyris, C. (1964), *Integrating the Individual and the Organisation*. New York: Wiley.

Argyris, C. (1971), "Management Information Systems: the Challenge to Rationality and Emotionality", *Management Science*, February.

Argyris, C. (1972), *The Applicability of Organisational Sociology*. Cambridge University Press.

Arrow, K.J. (1951), *Social Choice and Individual Values*. New York: Wiley.

Ashby, W.R. (1952), *Design for a Brain*. New York: Chapman & Hall.

Babbage, C. (1835), *On the Economy of Machinery and Manufactures*. London: Charles Knight.

Bailey, F.G. (1971), *Gifts and Poison*. London: Blackwell.

Baker, S. and R. Hansen (1975), "Job Design and Worker Satisfaction: a challenge to assumptions", *Journal of Occupational Psychology*, vol. 48, 79-91.

Barnard, C.I. (1938), *The Functions of the Executive*. Harvard University Press.

Barnett, J.H. (1970), "Individual Goals and Organisational Objectives: A study of integrative mechanisms". Institute for Social Research, University of Michigan.

Bateson, G., "Cultural determinants of personality", in J. Hunt (ed.) (1944), *Personality and the Behaviour Disorders*. New York: Ronald Press.

Bateson, G. (1972), *Steps to an Ecology of Mind*. New York: Paladin.

Bauer, R., A. Iukeles, C. Kluckhohn (1959), *How the Soviet System Works*. Harvard University Press.

Bauer, R. (1966), "Social Psychology and the Study of Policy Formulation", *The American Psychologist*, 903-912.

Beer, S. (1966), *Decision and Control*. New York: Wiley.

Beer, S. (1969), "The Aborting Corporate Plan", in E. Jantsch (ed.), *Perspectives of Planning*. Paris: OECD.

Beer, S. (1970), *The Organisation of the Manchester Business School*. Manchester Business School.

Beer, S. (1972), *The Brain of the Firm*. London: Allen Lane, The Penguin Press.

Beer, S. (1974), "Managing Modern Complexity", *Computers and People,* May, 18-21.

Bell, D. (1974), *The Coming of Post-Industrial Society.* New York: Heinemann.

Ben, David J. (1971), *The Scientist's Role in Society.* Englewood Cliffs, N.Y.: Prentice Hall.

Bennis, W.G. (1966), *Changing Organisations.* New York: McGraw Hill, 1960.

Bentham, J. (1907), *An Introduction to the Principles of Morals and Legislation.* Oxford University Press.

Bjorn Andersen, N., B. Hedberg (1977), "Designing information systems in an organisational perspective", *North Holland/TIMS Studies in the Management Sciences,* vol. 5, 125-142.

Black, M. (ed.) (1964), *Philosophical Analysis.* New York: J. Huca.

Blake, R.N., J. Mouton (1964), *The Managerial Grid.* Houston: Gulf.

Blau, P.M. (1964), *Exchange and Power in Social Life.* New York: Wiley.

Blau, P.M., W.R. Scott (1963), *Formal Organisations.* London: Routledge & Kegan Paul.

Blauner, R. (1960), "Work Satisfaction and Industrial Trends", in W. Galenson and S.M. Lipset (eds.), *Labour and Trade Unionism: an Interdisciplinary Reader.* New York: Wiley.

Blauner, R. (1964), *Alienation and Freedom: the Factory Worker and his Industry.* University of Chicago Press.

Boguslaw, R. (1965), *The New Utopians.* Englewood Cliffs, N.Y.: Prentice Hall.

Bostrom, R.P., J.S. Heinen (1977), "MIS problems and failures: a socio-technical perspective", *MIS Quarterly,* September, 17-32.

Bourke, V.J. (1970), *History of Ethics.* New York: Doubleday (image edition).

Bredemeier, H.C., R.M. Stephenson (1962), *The Analysis of Social Systems.* New York: Holt, Rinehard and Winston.

Brown, W. (1960), *Exploration in Management.* London: Heinemann.

Burden, D. (1976), "New model office project in the Department of Health and Social Security", *Management Services in Government,* vol. 31 (1).

Burns, T., G.M. Stalker (1961), *The Management of Innovation.* London: Tavistock.

Byrne, D., G.L. Clore (1967), "Effective arousal and attraction", *Journal of Personality and Social Psychology,* vol. 6 (4), no. 638, Monograph Supplement.

Campbell, D.T. (1961), "Conformity in psychology's theories of acquired behavioural dispositions", in I.A. Berg and B.M. Bass (eds.), *Conformity and Deviation.* New York: Harper.

Caplow, T. (1964), *The Sociology of Work.* New York: McGraw Hill.

Coser, L.A. (1965), *The Functions of Social Conflict.* Glenoe, Ill.: The Free Press.

Carroll, J. (1974), *Break-Out from the Crystal Palace.* London: Routledge and Kegan Paul.

Castaneda, C. (1968), *The Teaching of Don Juan.* University of California Press.

Cherns, A.B. (1976), "The principles of socio-technical design", *Human Relations,* vol. 29 (8), 783-792.

Christensen, S. (1974), *Decision Making and Ideology,* Forsknigsrapport 73-4. Copenhagen, Denmark: Institute of Organisation and Industrial Sociology.

Churchman, C. West (1968), *The Systems Approach.* New York: Delacourt Press.

Churchman, C. West (1968), "The case against planning", *Management Decision,* Summer.

Churchman, C. West (1971), *The Design of Inquiring Systems.* New York: Basic Books.

Churchman, C. West, A.M. Schainblatt (1965), "The researcher and the manager: a dialectic of implementation", *Management Science,* vol. 11 (4), 69-87.

Cohen, M.D., J.G. March, J.P. Olsen (1972), "A garbage can model of organisational choice", *Administrative Science Quarterly,* vol. 17 (1), 1-25.

Cooper, R. (1973), "How jobs motivate", *Personnel Review,* vol. 2, Spring.

Cooper, R. (1974), *Job Motivation and Job Design.* London: Institute of Personnel Management.

Coser, L.A., B. Rosenberg (eds.) (1957), *Sociological Theory.* New York: Macmillan.

Crozier, M. (1964), *The Bureaucreatic Phenomenon.* London: Tavistock Publication.

Dahrendorf, R. (1959), *Class and Class Conflict in Industrial Society.* London: Routledge and Kegan Paul.

Dale, R. (1962), *Planning and Developing the Company Organisation Structure.* A.M.A.

Dalton, M. (1959), *Men Who Manage.* New York: Wiley.

Davis, L.E. (1971), "Job satisfaction research: the post-industrial view", *Industrial Relations,* vol. 10, 176-193.

Davis, L.E., A.B. Cherns (1975), *The Quality of Working Life,* vols. 1 and 2. New York: The Free Press.

Davis, L.E., J.C. Taylor (1972), *Design of Jobs.* London: Penguin Books.

Dewey, J. (1916), *Democracy and Education.* New York: Macmillan.

Dewey, J. (1957), *Human Nature and Conduct: an Introduction to Social Psychology.* New York: The Modern Library.

Downie, R.S. (1971), *Roles and Values.* London: Methuen.

Drucker, P. (1962), *Congressional Testimony on Automation: Implications for the Future.* Morris Philopson (ed.). New York: Random House.

Duncan, R.B. (1973), "Multiple Decision-making structures in adapting to environmental uncertainty: the impact of organisational effectiveness", *Human Relations,* vol. 26 (3), 273-29.

Edstrom, A., L. Nauges (1974), "The implementation of computer based information systems under varying structural conditions". Paper presented to Altorg Conference, Gothenburg, May 1974.

Elliot, A.G.P. (1955), *Revising a merit rating scheme.* London: Institute of Personnel Management.

Ellul, J. (1967), *The Technological Society.* New York: Vintage Books.

Emerson, R.M. (1962), "Power Dependence Relations", *American Sociological Review,* vol. 27, 31-41.

Emery, F.E., E. Thorsrud (1969), *Form and Content in Industrial Democracy.* London: Tavistock.

Etzioni, A. (1968), *The Active Society*. London: Collier-Macmillan.

Etzioni, A. (1968), *A Theory of Societal and Political Processes*. New York: Free Press.

Evans, C. (1979), *The Mighty Micro*. New York: Gollancz.

Feigl, F. "The principii's non disputandum", in M. Black (ed.) (1954), *Philosophical Analysis*. New York: Ithica.

Festinger, L. (1957), *A Theory of Cognitive Dissonance*. Evanston, Ill.: Row Peterson.

Fiedler, F.E. (1967), *A Theory of Leadership Effectiveness*. New York: McGraw Hill.

Fitzgerald, T.H. (1971), "Why motivation theory doesn't work", *Harvard Business Review*, July-August.

Follett, Mary Parker (1924), *Creative Experience*. London: Longman.

Fox, A. (1974), *Beyond Contract: Work, Power and Trust Relations*. London: Faber.

Freire, P. (1972), *Pedagogy of the Oppressed*. London: Penguin.

French, S. (1953), *Civilisation and its discontents*. London: Hogarth.

Friedmann, J. (0000), "A conceptual model for the analysis of planning behaviour", *Administrative Science Quarterly*, vol. 12 (2), 225-252.

Fromm, E. (1968), *The Revolution of Hope*. New York: Bantam Books.

Gerstl, J.E., S.P. Hutton (1966), *Engineers: the Autonomy of a Profession*. London: Tavistock.

Gluckman, M. (1955), *Custom and Conflict in Africa*. Glencoe, Ill.: The Free Press.

Golembiewski, R.T. (1970), "Organisational properties and managerial learning: testing alternative models of attitude change", *Academy of Management Journal*, vol. 13 (1), 18-28.

Gouldner, A. (1955), *Patterns of Industrial Bureaucracy*. London: Routledge and Kegan Paul.

Gouldner, A.W. (1971), *The Coming Crisis of Western Sociology*. London: Heinemann.

Gowler, D., K. Legge (1975), "Occupational role integration and the retention of Labour", in B. Pettman (ed.), *Labour Turnover and Retention*. London: Halstead Press.

Gowler, D., K. Legge (1972), "Occupational role development, Part One", *Personnel Review*, vol. 1 (2), 12-27.

Gowler, D., "The concept of the person", in R. Rudduck (ed.) (1973), *Six Approaches to the Person*. London: Routledge and Kegan Paul.

Gowler, D. (1974), "Values, Contracts and Job Satisfaction", *Personnel Review*, vol. 3 (2), 4-15.

Gross, B.M. (ed.) (1974), *Action Under Planning*. New York: McGraw Hill.

Gross, L. (1967), *Sociological Theory: Inquiries and Paradigms*. New York: Harper and Row.

Hedberg, B. (1975), "Computer Systems to Support Industrial Democracy", in E. Mumford and H. Sackman (eds.), *Human Choice and Computers*. Amsterdam: North Holland.

Hedberg, B. (1978), "The Information Systems Design Process" (to be published shortly).

Hedberg, B., S. Jonsson (1976), "Strategy Formulation as a discontinuous process", F.E. Rapport 61, University of Gothenburg.

Hedberg, B., E. Mumford (1975), "The design of computer systems: a man's vision of man as an integral part of the systems design process", in E. Mumford and H. Sackman (eds.), *Human Choice and Computers*. Amsterdam: North Holland.

Hedberg, B., P. Nystrom, W. Starbuck (1976), "Camping on see-saws, prescription for a self-designing organisation", *Administrative Science Quarterly*, vol. 21 (1), 41-62.

Hekimian, J.S., H. Mintzberg (1968), "The planning dilemma: there is a way out", *Management Review*, May, 4-17.

Herbst, P. (1974), *Socio-technical design*. London: Tavistock.

Herzberg, F. (1966), *Work and the Nature of Man*. London: Staples Press.

Hickson, D.J., C.R. Hinings, C.A. Lee, R.E. Schneck, J.M. Jennings (1971), "A Strategic Contingencies Theory of Intra-Organisational Power", *Administrative Science Quarterly*, vol. 16 (2), 216-229.

Hitch, C.J. (1960), *On the Choice of Objectives in Systems Studies*. Monica, Calif.: Rand.

Hobson, Sir O. (1962), *The Post Office Savings Bank 1861-1961*. London: Central Office of Information.

Hollander, E.P., R.H. Willis (1967), "Some current issues in the psychology of conformity and non-conformity", *Psychological Bulletin*, vol. 68, 62-76.

Hunter, J., D.B. Crampton (1977), "Bridging the information gap", *Management Services in Government*, vol. 32 (3).

Hunt, J. (ed.) (1944), *Personality and the Behaviour Disorders*. New York: Ronald Press.

Jacques, E. (1961), *Equitable Payment*. London: Heinemann.

Jenkins, C. (1979), *The Collapse of Work*. London: Eyre Methuen.

Kahn, R.L., D. Wolfe, R.P. Quinn, J.D. Snoek, R.A. Rosenthal (1964), *Organisational Stress*. New York: Wiley.

Katz, D., B.S. Georgopoulos (1971), "Organisations in a chaning world", *Applied Behavioural Science*, vol. 7, no. 3, 342-369.

Katz, J. (1969), "Essences as moral identities: verifiability and responsibility in imputations of deviance and charisma", *American Journal of Sociology*, vol. 80 (6), 1369-1389.

Katzell, R.A., D. Yankelovich (1975), *Work Productivity and Job Satisfaction*. New York: The Psychological Corporation.

Kelly, G.A. (1963), *A Theory of Personality*. New York: Norton.

Kluckhohn, C. and others, "Value and value-orientations in the theory of action: an exploration in definition and classification", in T. Parsons and E. Shils (1951), *Toward a General Theory of Action*. Harvard University Press.

Knight, K.E. (1967), "A descriptive model of the intra-firm innovation process", *Journal of Business*, vol. 40, 5, 478-496.

Kuhn, T.S. (1962), *The Structure of Scientific Revolutions*. The University of Chicago Press.

Lamont, W.D. (1955), *The Value Judgement.* Edinburgh University Press.

Lawrence, P., J. Lorsch (1967), *Organisations and Environment.* Harvard University Press.

Leary, T. (1970), *The Politics of Ecstasy.* New York: Paladin.

Leavitt, H. (1958), *Managerial Psychology.* University of Chicago Press.

Lee, D. (1948), "Are basic needs ultimate?", *Journal of Abnormal and Social Psychology,* vol. XIII, 391-395.

Levinson, H. (1974), "Asinine attitudes towards motivation", *Harvard Business Review,* January-February, 70-76.

Likert, R. (1967), *The Human Organisation.* New York: McGraw Hill.

Lindholm, R., J.P. Norstedt (1975), *The Volvo Report.* Swedish Employers' Federation.

Locke, E.A. (1969), "What is job satisfaction?", *Organisational Behaviour and Human Performance,* vol. 4, 309-336.

Locke, E.A. (1970), "Job satisfaction and job performance: a theoretical analysis", *Organisational Behaviour and Human Performance,* vol. 5, 484-500.

Lucas, H. (1975), *Why Information Systems Fail.* Columbia University Press.

Lupton, T., D. Gowler (1969), *Selecting a Wage Payment System.* London: Engineering Employers' Federation.

Lystad, M.H. (1972), "Social Alienation: a review of current literature", *The Sociological Quarterly,* vol. 13, 90-113.

McCaskey, M.B. (1974), "A contingency approach to planning: planning with goals and planning without goals", *Academy of Management Journal,* vol. 17 (2), 281-289.

McGregor, D. (1960), *The Human Side of the Enterprise.* New York: McGraw Hill.

Mann, F.C., L.K. Williams (1960), "Observations on the dynamics of a change to electronic data processing equipment", *Administrative Science Quarterly,* vol. 5 (2), 217-256.

March, J.G., M.A. Simon (1958), *Organisations.* New York: Wiley.

Martin, J., A. Norman (1970), *The Computerised Society.* Englewood Cliffs, N.J.: Prentice Hall.

Marx, K. (1844), *The Economic and Philosophic Manuscripts of 1844,* D.J. Struik (ed.) (1965). New York: International Publishing Co.

Marx, K. (1862), *Das Kapital.* Trans. E. and C. Paul (1933). London: Dent.

Maslow, A.H. (1954), *Motivation and Personality.* New York: Harper.

Maslow, A.H. (1959), *New Knowledge in Human Values.* New York: Harper.

Mason, R.O. (1969), "A dialectic approach to strategic planning", *Management Science,* vol. 15, no. 8, 403-414.

Mason, R.O., I. Mitroff (1973), "A program for research and management information systems", *Management Science,* vol. 19, 475-487.

Mechanic, D. (1962), "Sources of power of lower participants in complex organisations", *Administrative Science Quarterly,* vol. 7 (3), 349-364.

Meehan, E.J. (1968), *Explanation in Social Science: a system paradigm.* Homewood, Ill.: Dorsey.

Meehan, E.J. (1969), *Value Judgement and Social Science.* New York: Dorsey.

Menzies, K. (1967), *Talcott Parsons and the Social Image of Man*. London: Routledge and Kegan Paul.

Miller, G.A., E. Galanter, K.M. Pribham (1960), *Plans and the structure of behaviour*. New York: Wiley.

Mills, C. Wright (1959), *The Sociological Imagination*. Oxford University Press.

Morley, E. (1974), "Human support systems in complex manufacturing organisations: a special case of differentiation", *Administrative Science Quarterly*, vol. 19 (3), 295-318.

Mumford, E. (1964), *Living with a computer*. London: Institute of Personnel Management.

Mumford, E. (1968), "Planning for computers", *Management Decision*, Summer 1968.

Mumford, E. (1969), "Implementing EDP Systems – a sociological perspective", *The Computer Bulletin*, January.

Mumford, E. (1969), *Computers, Planning and Personnel Management*. London: Institute of Personnel Management.

Mumford, E. (1970), "The Human Factor", *Data Processing*, July-August.

Mumford, E. (1970), "A new approach to using an old theory", *Sociological Review*, vol. 18 (1), 71-101.

Mumford, E. (1971), *Systems Design for People*. Manchester: National Computer Centre.

Mumford, E. (1971), "A Comprehensive Method for Handling the Human Factor when Introducing Computer Systems", *LFIP Conference Proceedings*, Holland.

Mumford, E. (1972), "Job Satisfaction: A Method of Analysis", *Personnel Review*, vol. 1 (3), 49-87.

Mumford, E. (1972), *Job Satisfaction: A Study of Computer Specialists*. London: Longmans.

Mumford, E. (1973), "Job Satisfaction: A Major Objective for the System Design Process", *Management Information*, vol. 2 (4), 191-202.

Mumford, E. (1974), "Computer Systems and Work Design: Problems of Philosophy and Vision", *Personnel Review*, vol. 3 (2), 40-49.

Mumford, E. (1976), "Computers and Democracy", *Personnel Management*, September.

Mumford, E. (1978), "Procedures for the Institution of Change in Work Organisation", in *Social Aspects of Work Organisation: Implications for Social Policy and Labour Relations*. Geneva: International Institute for Labour Studies, Research Series, No. 33.

Mumford, E. (1978), "The Design of Work: New Approaches and New Needs". To be published in the Proceedings of IFAC, 1977.

Mumford, E., O. Banks (1967), *The Computer and the Clerk*. London: Routledge and Kegan Paul.

Mumford, E., F. Land, J. Hawgood (1978), "A Participative Approach to Planning and Designing Computer Systems and Procedures to Assist This", *Impact of Science on Society*, vol. 28 (3).

308

Mumford, E., D. Mercer, S. Mills, M. Weir (1972), "The Human Problems of Computer Introduction", *Management Decision*, vol. 10 (1), 6-17.

Mumford, E., D. Mercer, S. Mills, M. Weir (1972), "Discuss the Suggestion that no Project Team Concerned with the Implementation of the Management Information System should be without a trained Social Scientist", *Computer Bulletin*, June 1972, 356-358.

Mumford, E., A. Pettigrew (1975), *Implementing Strategic Decisions*. London: Longman.

Mumford, E., H. Sackman (eds.) (1975), *Human Choice and Computers*. Amsterdam: North Holland.

Mumford, E., T.B. Ward (1966), "Computer Technologists: Dilemmas of a New Role", *Journal of Management Studies*, vol. 3, 244-255.

Mumford, E., T.B. Ward (1966), "The Hard Facts About the Software Merchants", *The Director*, vol. 18, 368-469.

Mumford, E., T.B. Ward (1968), "Computer People and their Impact", *New Society*, September 1968.

Mumford, E., T.B. Ward (1968), *Computers: Planning for People*. London: Batsfords.

Mumford, L. (1964), *The Pentagon of Power*. London: Secker and Warburg.

Najder, Z. (1975), *Values and Evaluations*. Oxford: Clarendon Press.

Neal, M.A. (1965), *Values and Interests in Social Change*. Englewood Cliffs, N.J.: Prentice Hall.

Nystrom, P.C. (1975), "Resisters, Adapters and Innovators: On the Implementation of Managerial Goal Setting Systems", working paper, University of Wisconsin, Milwaukee.

Osgood, C.E. (1964), *Method and Theory in Experimental Psychology*. Oxford University Press.

Ozbekhan, H. (1969), "Toward a General Theory of Planning", in E. Jantsch (ed.), *Perspectives of Planning*. Paris: OECD, 1969.

Pace, D.E. (1975), "A Job Satisfaction Approach to a Management Problem", *Management Services in Government*, vol. 30 (4).

Parsons, T. (1961), "Polarization and the Problem of International Order", *Berkeley Journal of Sociology*, vol. 6 (1), 115-134.

Parsons, T. (1964), *Essays in Sociological Theory*. New York: The Free Press.

Parsons, T. (1964), *The Social System*. New York: The Free Press.

Parsons, T. (1964a), "Malinowski and the Theory of Social Systems", in R. Firth (ed.), *Man and Culture: An Evaluation of the Work of Bronislaw Malinowski*. New York: Harper and Row.

Parsons, T. (1965b), "The Point of View of the Author", in M. Black (ed.), *The Social Theories of Talcott Parsons: A Critical Examination*. Englewood Cliffs, N.J.: Prentice Hall.

Parsons, T., E. Shils (1951), *Towards a General Theory of Action*. Harvard University Press.

Parsons, T., E. Shils, K.D. Naegle, J.E. Pitts (eds.) (1968), *Theories of Society*. New York: The Free Press.

Peattie, L.R. (1965), "Anthropology and the Search for Values", *Journal of Applied Behavioural Science,* vol. 1 (4), 361-368.

Pennings, J.M. (1970), "Work-value Systems of White Collar Workers", *Administrative Science Quarterly,* vol. 1 (4), 361-368.

Perry, R.B. (1926), *General Theory of Value.* Harvard University Press.

Pettigrew, A. (1968), "Inter-group Conflict and Role Strain", *Journal of Management Studies,* vol. 5 (2), 205-218.

Pettigrew, A. (1973a), "Occupational Specialisation as an Emergent Process", *The Sociological Review,* vol. 21 (2), 255-278.

Pettigrew, A. (1973), *The Politics of Organizational Decision Making.* London: Tavistock Publications.

Philips, D. (1973), *Abandoning Method.* New York: Jossey-Bacy.

Plato, *The Republic,* Book IV.

Podger, D. (1978), unpublished Ph.D. thesis.

Popper, K. (1963), *The Open Society and its Enemies,* vol. 1. Princeton University Press.

Prandy, K. (1965), *Professional Employees.* London: Faber.

Quinn, R., T. Mangiones, M.S. Baldi de Mandilovitch (1973), "Evaluating Working Conditions in America", *Monthly Labor Review,* November 1973, 32-41.

Rescher, N. (1969), *Introduction to Value Theory.* Englewood Cliffs, N.J.: Prentice Hall.

Rittel, W.J., R.M. Webber (1973), "Dilemmas in a General Theory of Planning", *Policy Sciences,* vol. 4, 155-169.

Robertson, K. (1973), "Job Satisfaction in the Civil Service – A Perspective", *Management Services in Government,* vol. 28, no. 3.

Rokeach, M. (1973), *The Nature of Human Values.* New York: the Free Press.

Rosenbrock, H. (1975), "The Future of Control". Paper presented to the 6th IFAC Congress, Boston.

Ross, W.D. (1930), *The Right and the Good.* Oxford: Clarendon Press.

Ruddock, R. (ed.) (1973), *Six Approaches to the Person.* London: Routledge and Kegan Paul.

Sackman, H. (1971), *Mass Information Utilities and Social Excellence.* New York: Auerbach.

Sackman, H. (1971), *Computers, Systems Science and Evolving Society.* New York: Wiley.

Saenger, G., S. Flowerman (1954), "Stereotypes and Prejudicial Attitudes", *Human Relations,* vol. 7, 217-240.

Salanik, G.R., G. Pfetter (1977), "Who Gets Power – and how they hold onto It: A Strategic-Contingency Model of Power", *Organisational Dynamics,* Winter, 3-21.

Sayles, L.R. (1974), "The Innovation Process: An Organizational Analysis", *Journal of Management Studies,* vol. 11 (3), 175-263.

Schein, E.H. (1965), *Organizational Psychology.* Englewood Cliffs, N.J.: Prentice Hall.

Schlick, M. (1939), *Problems of Ethics*. New York: Prentice Hall.

Schon, D.A. (1967), *Technology and Change*. London: Pergamon.

Schon, D.A. (1971), *Beyond the Stable State*. New York: Random House.

Seashore, S.E. (1973), "Job Satisfaction: A Dynamic Predictor of Adaptive and Defensive Behaviour", *Studies in Personnel Psychology*, vol. 5 (1), 7-20.

Seashore, S.E., T.D. Taber (1975), "Job Satisfaction Indicators and their Correlates", *American Behavioural Scientist*, vol. 18 (3), 333-369.

Selznick, P. (1969), *Law, Society and Industrial Justice*. New York: Russell Sage Foundation.

Shephard, H.A. (1965), "Changing Interpersonal and Intergroup Relationships in Organizations", in J.G. March (ed.), *Handbook of Organization*. Chicago: Rand McNally.

Sherif, M. (1966), *The Psychology of Social Norms*. New York: Harper.

Silverman, D. (1975), *Reading Castaneda*. London: Routledge and Kegan Paul.

Singer, E.A. (1948), *In Search of a Way of Life*. Columbia University Press.

Slater, P. (1971), *The Pursuit of Loneliness*. New York: Beacon Press.

Smelser, N. (1959), *Social Change in the Industrial Revolution*. London: Routledge and Kegan Paul.

Smith, A. (1910), *The Wealth of Nations*. London: Dent.

Smith, A. (1975), "The Functions of Work", *Omega*, vol. 3 (4), 383-393.

Steiner, G. (1969), *Top Management Planning*. New York: Macmillan.

Stone, E.E., L. Parker (1975), "Job Characteristics and Job Attitudes: A Multivariate Study", *Journal of Applied Psychology*, vol. 60 (1), 57-64.

Strivastra, S. (1975), *Job Satisfaction and Productivity*. Cleveland: Dept. of Organizational Behaviour, Case Western Reserve University.

Swanson, B. (1974), "Systems Heros", *General Systems*, vol. 19, 91-95.

Sykes, A.J.M. (1964), "A Study in Changing the Attitudes and Stereotypes of Industrial Workers", *Human Relations*, vol. 17, 143-154.

Taylor, F.W. (1947), *Scientific Management*. New York: Harper and Row.

Taylor, J.C. (1975), "The Human Side of Work: The Socio-Technical Approach to Work Systems Design", *Personnel Review*, vol. 4.

Taylor, J.C. (1976), *Job Satisfaction and Quality of Working Life: A Reassessment*. California: Graduate School of Management UCLA.

Tawney, R. (1938), *Religion and the Rise of Capitalism*. London: Harmondsworth, Penguin.

Thomas, W.I., "The Unadjusted Girl", in L.A. Coser and B. Rosenberg (eds.) (1957), *Sociological Theory*. New York: Macmillan.

Thompson, E.P. (1968), *The Making of the English Working Class*. London: harmondsworth, Penguin.

Toffler, A. (1971), *Future Shock*. New York: Pan Books.

Touraine, A. (1955), "L'évolution du travaillaux usines Renault", in C.R. Walker (ed.) (1962), *Modern Technology and Civilisation*. McGraw Hill.

Touraine, A. (1974), *The Post-Industrial Society*. London: Wildwood House.

Trist, E., K.W. Bamforth (1951), "Some Social and Psychological Consequences of the Longwall Method of Coal-Getting", *Human Relations*, vol. 4, no. 1, 3-38.

Turner, A.N., P.R. Lawrence (1965), *Industrial Jobs and the Worker: An Investigation of Response to Task Attributes*. Boston: Harvard University Press.

Vickers, G. (1968), *The Art of Judgement*. New York: Basic Books.

Vickers, G. (1973), *Making Institutions Work*. London: Associated Business Programmes.

Vroom, V. (1960), *Some Personality Determinants of the Effects of Participation*. Englewood Cliffs, N.J.: Prentice Hall.

Vroom, V. (1964), *Work and Motivation*. New York: Wiley.

Wanores, J., E. Lawler (1972), "Measurement and Meaning of Job Satisfaction", *Journal of Applied Psychology*, vol. 56 (2), 95-105.

Ward, T. (1973), *Computer Organization, Personnel and Control*. London: Longman.

Wedderburn, D., K. Crompton (1972), *Workers' Attitudes and Technology*. Cambridge University Press.

Weick, K. (1969), *The Social Psychology of Organizing*. Reading, Mass.: Addison Wesley.

Whisler, T.L. (1970), *The Impact of Computers in Organizations*. New York: Praeger.

Weizenbaum, J. (1976), *Computer Power and Reason*. New York: Freeman.

White, J.K., R.A. Ruh (1973), "The Effects of Personal Values on the Relationship between Participation and Job Attitudes," *Administrative Science Quarterly*, vol. 18 (4), 506-514.

White, R.W. (1959), "Motivation Reconsidered: The Concept of Competence", *Psychological Review*, vol. 66, 295-333.

Author Index

Ackoff and Rivett (1963), 236
Allport (1935), 242; (1954), 243
Andersen (1977), 109
Andersen and Hedberg (1977), 234-5
Argyris (1964), 269; (1971), 63; (1972), 238
Arrow (1951), 13
Ashby (1952), 240

Bailey (1971), 17
Baker and Hansen (1975), 269
Barnard (1938), 240
Barrett (1970), 269
Bateson (1972), 9
Bauer (1959), 244; (1966), 233-4
Beer (1966), 10; (1969), 94, 231; (1970), 53; (1972), 4, 236, 240; (1974), 284
Beer and Hedberg (1974 and 1975), 238
Bell (1974), 277
Ben-David (1971), 243
Bennis (1966), 231, 243
Bentham, Jeremy, 5
Black (1964), 21
Blake and Mouton (1964), 41
Blau, 243
Blau and Scott (1963), 263
Blauer (1964), 274
Boguslaw (1965), 61
Bostrom and Heinen (1975 and 1977), 264
Bourke (1968), 21; (1970), 9
Bredemeier (1962), 94, 263
Burden (1976), 276
Burns and Stalker (1961), 244
Byrne and Clore (1967), 262

Campbell (1961), 237
Caplow (1964), 241
Carroll (1974), 5-6

Castaneda (1968), 11
Cherns (1976), 149
Christensen (1971), 251
Churchman (1965), 227; (1968), 241; (1971), 63
Cohen *et al.* (1972), 250
Cooper (1973), 53-4; (1974), 42, 54
Crozier (1964), 41, 237

Dahrendorf (1959), 12, 241
Dale (1962), 231
Dalton (1959), 237
Davis (1971), 7, 61, 267; (1972), 42, 53, 149; (1975), 149
Dewey (1957), 254
Downie (1971), 242
Drucker (1962), 278
Duncan (1972), 240

Edstrom (1974), 258, 261
Ellul (1967), 21
Emerson (1962), 243
Emery (1969), 42
Etzioli (1968), 236, 240, 248, 250
Evans (1969), 270; (1979), 279

Fiedler (1967), 41
Fitzgerald (1971), 263
Follet (1924), 282, 284
Fox (1974), 7, 265, 274, 277-8
Freier (1972), 8
Friedman (1967), 251, 256-7, 259
Fromm (1968), 60

Gerstl and Hutton (1966), 243
Gluckman (1955), 241
Golembiewski (1970), 254
Gouldner (1955), 41; (1971), 265

Subject Index

Aims and objectives of this study will be found in the entry for Study. The letter-by-letter system has been adopted.

COLOPHON

letter: Baskerville 10/11, 9/10, Univers
setter: Expertext, Alphen aan den Rijn
printer: Samsom-Sijthoff Grafische Bedrijven, Alphen aan den Rijn
binder: Callenbach, Nijkerk
cover design: Wim Bottenheft

318